David C. Brown
AI Research Group
Worcester Polytechnic Institute

B. Chandrasekaran
Laboratory for AI Research
Ohio State University

Design Problem Solving
Knowledge Structures
and Control Strategies

Pitman, London
Morgan Kaufmann Publishers, Inc., San Mateo, California

PITMAN PUBLISHING
128 Long Acre, London WC2E 9AN

A Division of Longman Group UK Limited

© David C. Brown and B. Chandrasekaran 1989

First published 1989

Available in the Western Hemisphere from
MORGAN KAUFMANN PUBLISHERS, INC.,
2929 Campus Drive, San Mateo, California 94403

ISSN 0268-7526

British Library Cataloguing in Publication Data
Brown, David C.
  Design problem solving : knowledge
  structures and control strategies.—
  (Research notes in artificial intelligence,
  ISSN 0268-7526)
  1. Expert systems (Computer science)
  2. Computer engineering
  I. Title II. Chandrasekaran, B. III. Series
  006.3'3   QA76.76.E95

ISBN 0 273 08766 5

Library of Congress Cataloging in Publication Data
Brown, David C., 1947–
  Design problem-solving.

  (Research notes in artificial intelligence,
ISSN 0268-7526)
  Bibliography: p.
  1. Expert systems (Computer science) 2. System
design. I. Chandrasekaran, B., 1942– . II. Title.
III. Series.
QA76.76.E95B77 1987   006.3'3   86-34418
  ISBN 0-934613-07-9 (Morgan Kaufmann)

All rights reserved; no part of this publication may be reproduced,
stored in a retrieval system, or transmitted in any form or by any
means, electronic, mechanical, photocopying, recording or
otherwise without either the prior written permission of the publishers
or a licence permitting restricted copying issued by the Copyright
Licencing Agency Ltd, 33–34 Alfred Place, London WC1E 7DP.
This book may not be lent, resold, hired out or otherwise disposed of
by way of trade in any form of binding or cover other than that in
which it is published, without the prior consent of the publishers.

Reproduced and printed by photolithography
in Great Britain by Biddles Ltd, Guildford

Contents

1 **Introduction** 1
   1.1 What is the Book About? 1
   1.2 Artificial Intelligence *vs.* Traditional Computational Methods 1
   1.3 Generic Tasks: Expert Systems Beyond Rules and Frames 6
      1.3.1 Characterization of a generic task 8
      1.3.2 Hierarchical classification 9
      1.3.3 High level languages based on generic tasks 10
      1.3.4 Critiques of the GT approach 11
      1.3.5 Task-specific architectures 14

2 **A Framework for Design Problem Solving** 19
   2.1 What is the Design Problem? 19
   2.2 What Kind of Space to Search? 21
   2.3 Information Processing Analysis of Design 23
      2.3.1 Processes that propose design choices 23
      2.3.2 Auxiliary processes 30
   2.4 Implications of Above Analysis 31
   2.5 Classes of Design 32
      2.5.1 Class 1 design 33
      2.5.2 Class 2 design 33
      2.5.3 Class 3 design 34
      2.5.4 A class 3 product 34
      2.5.5 Class 3 complexity 35
      2.5.6 Imprecision of the classification 35

3 **Expert System Architecture for Class 3 Design** 37
   3.1 The Structure of a Class 3 Design Problem Solver 37
   3.2 Design Agents 37
      3.2.1 Specialists 37
      3.2.2 Plans 39
      3.2.3 Tasks 41
      3.2.4 Steps 42
      3.2.5 Constraints 44
      3.2.6 Types of design knowledge 45
      3.2.7 Dependency and dependency measures 46
   3.3 DDB: The Design Database 47
      3.3.1 Drawings 47
      3.3.2 An hypothesis 47
      3.3.3 Alterations and revisions 48
      3.3.4 DDB constraints 48
   3.4 Other Agents 49

3.5 The Action of a Class 3 Design Problem Solver   49
    3.5.1 Requirements checking   50
    3.5.2 Rough-design   50
    3.5.3 Design   51
    3.5.4 Redesign and re-design   51
3.6 Inter-agent Communication   51
    3.6.1 Types of messages   52
3.7 Design Agent Action   53
    3.7.1 Specialist action   53
    3.7.2 Task action   55
    3.7.3 Step action   55
    3.7.4 Constraint action   55
3.8 Plan Selection   56
    3.8.1 The selection process   56
    3.8.2 Qualities of plans   60
    3.8.3 Situation factors   62
    3.8.4 Plan complexity   63
3.9 Summary   64

4 Failure Handling in Routine Design   65

4.1 Introduction to Failure Handling   65
    4.1.1 Restricted knowledge   65
    4.1.2 Social metaphor   66
    4.1.3 Local decisions   66
    4.1.4 Domain driven   66
4.2 Knowledge for Failure Handling   67
    4.2.1 Failure handlers   67
    4.2.2 Recovery from failure   69
4.3 Redesign Problem Solving   71
    4.3.1 How redesign occurs   71
    4.3.2 Design *vs.* redesign *vs.* re-design   72
4.4 An Overview of Failure Handling   72
4.5 Design Agent Failure   73
    4.5.1 Constraint failure   73
    4.5.2 Failure in a step   74
    4.5.3 Failure in a task   76
    4.5.4 Failure in a specialist   81
4.6 Failure Handling in Other Research   87
    4.6.1 Styles of failure handling   87
    4.6.2 EL/ARS   88
    4.6.3 TROPIC   89
    4.6.4 DESI/NASL   89
    4.6.5 BUILD   89
    4.6.6 MEND   90
    4.6.7 Other research   90
4.7 Summary   91

5 DSPL: A Language for Design Expert Systems 93
- 5.1 Introduction 93
  - 5.1.1 The design specialists and plans language 93
  - 5.1.2 DSPL conventions 94
- 5.2 Specialist Example 94
- 5.3 Plan Example 95
- 5.4 Task Example 96
- 5.5 Step Example 96
- 5.6 Constraint Example 98
- 5.7 Redesigner Example 99
- 5.8 Failure Handler Example 101
- 5.9 Sponsor Example 102
- 5.10 Selector Example 102
- 5.11 Summary 105

6 AIR-CYL: An Air Cylinder Design System 107
- 6.1 An Instance of Class 3 design 107
  - 6.1.1 The air cylinder 107
  - 6.1.2 The conceptual structure for the air cylinder 108
- 6.2 The AIR-CYL System 110
  - 6.2.1 Requirements checking 110
  - 6.2.2 Rough design 110
  - 6.2.3 Design 111
  - 6.2.4 Plan selection 112
  - 6.2.5 Failure handlers 113
  - 6.2.6 Redesigners 113
- 6.3 The DSPL System 114
  - 6.3.1 System setup 114
  - 6.3.2 Attribute tables 114
  - 6.3.3 Standard measures 115
  - 6.3.4 Tolerances 115
  - 6.3.5 Messages 115
  - 6.3.6 The DSPL interpreter 116
- 6.4 The Design Database 116
  - 6.4.1 The frame hierarchy 116
  - 6.4.2 Changes and updates 117
  - 6.4.3 Performance 119
- 6.5 Summary 120

7 Design Problem Solving: A Research Agenda 121
- 7.1 Improvements to DSPL System Support 121
  - 7.1.1 Interfaces 121
  - 7.1.2 Design system builder's aids 122
- 7.2 Problem solving in DSPL 123
  - 7.2.1 Conjunctive suggestions 123
  - 7.2.2 Plan selection 124
  - 7.2.3 Agent memory 124
  - 7.2.4 Multiple designs 124

       7.2.5   Failure-handling strategies   125
       7.2.6   Evaluating the design   125
       7.2.7   Post-design re-design   126
       7.2.8   Rough design   126
       7.2.9   Relaxation of requirements   127
       7.2.10 Automatic construction   127
       7.2.11 Performance degradation   128
       7.2.12 Use of defaults and catalogs   128
       7.2.13 Evaluation of the system   128
7.3   Limitations of Class 3 Approach   129
7.4   Directions for Design Research   130
       7.4.1   Investigating compilation of design knowledge   130
       7.4.2   Failure analysis   131
       7.4.3   Decomposition   131
       7.4.4   Design by weak methods   131
7.5   Conclusions   132

References   135

Appendix A: **Design Trace**   145

Appendix B: **Design Trace with Step Redesign**   153

Appendix C: **Design Trace with Task Redesign**   171

Appendix D: **Plan Selection Trace**   181

Appendix E: **DSPL Syntax**   191

List of Figures

Figure 1 Relation Between Knowledge at Structure-Function, Generic Task, and Higher Order Task Levels   13
Figure 2 A Specialist   38
Figure 3 Specialist "Head"   39
Figure 4 Plan "HeadDP1"   40
Figure 5 Task "AirCavity"   41
Figure 6 A Task   42
Figure 7 Step "AirCavityID"   43
Figure 8 Constraint "ACID"   45
Figure 9 Plan Refinement   54
Figure 10 Plan Selection   57
Figure 11 Sponsor "ExampleDPSponsor"   58
Figure 12 Selector "ExampleDPSelector"   59
Figure 13 Failure Handlers   68
Figure 14 Step Redesigner Structure   69
Figure 15 The Step's Failure Handler   75
Figure 16 The Step's Redesign Action   76
Figure 17 The Task Backup Strategy   78
Figure 18 Redesign Request of a Task   84
Figure 19 Redesign Request of a Specialist   85
Figure 20 Specialist "Head"   94
Figure 21 Plan "HeadDP1"   95
Figure 22 Task "AirCavity"   96
Figure 23 Step "AirCavityID"   97
Figure 24 Constraint "ACID"   98
Figure 25 Redesigner "AirCavityIDRedesigner"   100
Figure 26 Failure Handler "SystemStepBodyFailureFH"   101
Figure 27 Sponsor "ExampleDPSponsor"   103
Figure 28 Selector "ExampleDPSelector"   104
Figure 29 Air-Cylinder   108
Figure 30 Design Trace   108
Figure 31 Partial AIR-CYL Structure   109
Figure 32 Rough Design Hierarchy   110
Figure 33 Design Hierarchy   111
Figure 34 The Frames of the Design Data-Base   117
Figure 35 Memory Utilization   119
Figure 36 Timing Figures   119

Preface and Acknowledgements

This book owes its origin to the Ph.D. dissertation by David C. Brown completed in August 1984, under the supervision of B. Chandrasekaran, that describes the application of the generic task methodology developed by the latter to the problem of routine design. Chapters 3 onwards reflect the work reported in the dissertation, updated appropriately. The analysis of design as an information processing activity (Chapter 2) is of more recent vintage, as is the description of the generic task methodology and the discussion of what constitutes the essence of the AI approach to problem solving (Chapter 1).

While the book as a whole reflects a collaboration between the authors, Chandrasekaran largely takes responsibility for Chapters 1 and 2 describing the basic design problem solving framework, while the description of the DSPL language and the AIR-CYL Air Cylinder design system are largely the work of Brown. In both the Laboratory for AI Research at Ohio State, which Chandrasekaran directs, and in the AI Research Group at Worcester Polytechnic Institute, which Brown directs, significant additional work on extending the view of design described in this book has taken place. However, due to a desire to stop somewhere and get the book in print, we have mainly attempted to describe the basics of our approach to design problem-solving, rather than include all of our most recent ideas on the subject.

This work was supported at the Ohio State University by Air Force Office of Scientific Research Grant AFOSR 82-0255. We would also like to acknowledge the cooperation of the Accuray Corporation, Columbus, Ohio, as well as that of Dave Herman and Pete Schmitz.

B. Chandrasekaran acknowledges the support of Defense Advanced Research Projects Agency contract, Rome Air Development Center, F30602-85-C-0010.

We would also like to acknowledge the help and stimulation provided by the members of the Laboratory for Artificial Intelligence Research at the Ohio State University, as well as the members of the Artificial Intelligence Research Group at Worcester Polytechnic Institute.

Pitman Publishing Ltd., and our editors, Professor D. Sleeman and Dr. N. S. Sridharan, should be thanked for their continued patience and support during the long period over which this book has been developed.

1 Introduction

1.1 What is the Book About?

This is a book about design problem solving. In particular it is about how to build knowledge-based expert systems that perform types of design that we characterize as a form of *routine design*. It is assumed that the reader is reasonably familiar with the idea of expert systems, particularly the ideas of knowledge representation and inference as the key issues in the design of knowledge-based systems.

Design is a complex activity, and a number of systems exist which solve interesting and complex design problems in different domains. In studying design as a problem solving activity, a number of different goals could be adopted. For example, one might wish to build a highly sophisticated design system in a specific domain, using any of the current AI techniques, such as rule-based programming. In contrast, our aim has been to produce a generic theory of one type of design, and support it with an *architecture* and *a high level language* that is suitable for that type of design.

How successful one considers this work depends on the extent that the type of design identified is sufficiently general and interesting, and whether the approach and the technology offered help to build design systems of this type faster, more correctly, more perspicuously, and so forth.

In addition, we have also attempted to offer a general framework for the analysis of design which identifies several of the subprocesses of design. We believe that this analysis makes it possible to extend the approach presented in this book to larger classes of design.

1.2 Artificial Intelligence vs Traditional Computational Methods

A brief discussion of what kinds of computations characterize AI may be useful to clear up a skepticism that the term "expert system" often engenders. This is especially true of persons from the engineering domain, where highly sophisticated and complex computational methods have long been in use.

Does AI describe a set of computations with a different character than the kind of computations with which engineers are familiar? Why isn't an algorithm that numerically solves a partial differential equation to give answers about stresses and temperatures in a reactor vessel an "expert" system? After all, those equations embody the knowledge of some domain expert. Is a linear programming algorithm that gives an optimal solution to a scheduling problem an expert system? Should it be rewritten in some AI language for it to qualify as an expert system? Is a Bayesian decision function for diagnosis an expert system for diagnosis? Why, or why not? If it is, does AI have a unique subject matter? After all, Bayesian decision functions have been a long-time staple of statistical decision theories, as have linear programming algorithms in operations research.

Some people take the programming technology used as the criterion for whether a system is an "expert" system or not. Use of rule-based or frame-based systems, or Prolog, during its construction is supposed to qualify a system as an AI or expert system. The fallacy in that argument is clear when one considers that these AI programming languages are like Turing-machines. Any algorithm can be implemented using these languages. Making a multiplication algorithm written in OPS5 an expert system removes all specific content from the subject of expert systems. The subject will then become the study of any and all algorithms.

The problem is that people within AI, as well as outside of it, have no consensus about what sort of computations characterize intelligence. Intuitively it is clear that people are quite poor at some kinds of information processing, while they seem to do other kinds quite well. Without paper and pencil or a calculator, we cannot execute a linear programming algorithm or multiply two 32-digit numbers, while we can make interesting hypotheses, consider alternative explanations, etc., fairly naturally and effortlessly. We can simulate certain physical phenomena qualitatively, but we cannot (again, without paper and pencil) give exact numerical values for predictions derived in that fashion.

These observations apply even when we are working in a domain where we might be an expert. Human experts engage in a behavior where "thinking" alternates with calculations using various means of computational assistance. For example, designers think of plausible designs, and if some equations need to be worked out to get some parameters correct, they use paper and pencil, calculators, or computer programs to get the numbers that they need.

What we need is an account of the kinds of computations that characterize (cognitive) intelligent behavior, so that we can understand what the subject matter of

AI is and how it differs from non-AI computations.

It is a working hypothesis in AI that computations which underlie intelligent cognitive behavior are *discrete, symbolic, and qualitative*. But this doesn't characterize intelligent computation sufficiently. On one hand, it does not characterize the sense in which intelligent computations differ from merely being Turing-computable functions, and on the other, the qualitativeness of intelligent computation might be taken to imply that intelligence merely obtains an approximate solution, while quantitative computations are the ones to use for getting correct answers. That is, emphasis on qualitativeness alone doesn't bring out the *power* of intelligence. It doesn't answer the question of why, even with a number of quantitative models available, it may still be important to employ methods that are qualitative in the way intelligence is qualitative.

In [Chandrasekaran 88a], one of us has proposed a view of what makes intelligent cognitive behavior a special kind of computation. Intelligent behavior is characterized by a collection of general strategies that use knowledge in such a way that the complexity of computation inherent in certain tasks is minimized. For example, we will see in the next chapter that the design problem can be viewed fundamentally as a search problem in a very large space. Each element of this space reflects a possible candidate for the answer to the design problem. Checking whether a candidate design meets the specifications may additionally involve specific quantitative or other algorithmic methods. Thus the total computational effort, the search in the design space plus the design evaluation, can become very large. This places a large premium on searching the design solution space efficiently, so that only plausible designs are subjected to the full rigors and computational expense of detailed analysis.

One way to clarify some of the issues is to make a distinction between computations which are *being intelligent* versus those which use the result of earlier intelligent behavior. One might look at an algorithm for the greatest common divisor, and exclaim, "What a clever algorithm!". In reality, the creator of the algorithm was the one who was being clever during its construction. The algorithm itself, during its running, is not engaging in any of the processes that intelligence is composed of, such as exploration of possibilities, hypothesis-making, etc., or using any of the general methods for such behavior.

We can focus the discussion by considering what it means to be intelligent in problem solving. MYCIN, R1/XCON, finite element methods, linear programming algorithms, multiplication algorithms, etc., are all computational methods which provide solutions to some problems. Let us now consider a subclass of methods which are "intelligent" in the following sense: they explore a *problem space*, implicitly defined by a *problem representation*, using general search strategies which exploit typically qualitative heuristic knowledge about the problem domain. A working hypothesis in AI is that humans, unassisted by other computational techniques or paper and pencil, engage in this kind of problem solving. Newell [Newell 80] has described this hypothesis in some detail and called it the *Problem Space Hypothesis*.

The subarea of AI concerned with problem solving takes as its subject matter the phenomena that surround this kind of knowledge-based and general strategy-directed exploration of problem spaces. The power of these phenomena come from the effective way in which they explore very large problem spaces to make plausible and interesting hypotheses, which can then be verified by other means if necessary. Also, if information that can only be obtained by other kinds of computations are necessary during this kind of exploratory problem solving activity, then these other methods can be invoked, much as an engineer flits between, on one hand, "thinking" about a problem and making intermediate hypotheses, and, on the other hand, writing down some equations to solve before engaging in further exploration.

For lack of a better term, let us call computational methods that are characterized by such knowledge-based and strategy-directed exploration of qualitative problem spaces *problem space exploratory* techniques. Other kinds of computational techniques, let us call *solution algorithms*. These terms are unsatisfactory, but with proper qualifications they will do.

Note that our distinction is somewhat different from the fairly classical "heuristic" vs "algorithmic" distinction. For example, all known algorithms for the Traveling Salesman problem are inefficient, so a number of programs which approximate the solution by making various assumptions and approximations have come to be called heuristic solutions to the problem. But, these are still solution algorithms according to our definition, albeit without the properties of provable optimality of the solutions given by them, since these algorithms do not, at run time, engage in exploration of the underlying problem space by use of explicit knowledge and general exploration strategies.

There are many problems for which solution algorithms which are not computationally complex are known. Computer science, as well as a number of other disciplines, take as their subject matter the production of solution algorithms for a number of problems of a general or domain-specific nature. Sorting, multiplication, and linear programming are of this type.

When problems of this type are identified in any domain, there is no reason to engage in problem-space exploratory techniques. Adopting AI-type solutions to these problems will merely produce solutions which are qualitative and approximate. In addition, in comparison with the corresponding solution algorithms, the AI methods are likely to be computationally expensive as well. If during diagnostic reasoning one needs to know the exact value of pressure in the reactor chamber, if one has an equation that can be evaluated for it, and if one has all the information needed to evaluate the equation, then it is silly to use problem-space exploratory methods. On the other hand, for a number of problems such as diagnosis and design in the general case, the underlying spaces can be very large, and solution algorithms of restricted complexity are generally not available. This is when AI method are appropriate.

It is important to emphasize that expert behavior consists of both kinds of computations. Thus expert systems should use both kinds of techniques, employing each kind where their power is needed. Many of these solution algorithms are domain-specific, or use methods, such as linear programming, that are not the subject matter of AI per se. Consequently, it is most useful to confine the discussion in AI to problem space exploratory techniques, especially those inspired by human intelligence. For example, this book studies human-like problem space exploration for the task of design. While we do not discuss use of any computations of the solution algorithm type, it is assumed that such methods ought to be used whenever possible during the solution of any of the design subproblems in a specific domain.

Regarding exploration strategies, we have identified a number of general strategies that we call *generic tasks* to set up and explore problem spaces. This generic task theory is the subject of the next section.

1.3 Generic Tasks: Expert Systems Beyond Rules and Frames

Expert system building has come to mean programming in some combination of rule-based, frame-based or logic-based systems. In particular, the idea that one acquires and represents domain knowledge in some general form using one or more of these languages, and uses various kinds of inference machinery to use the knowledge so acquired to solve problems has virtually become second nature to system builders.

Our approach to knowledge-based systems is based on the view that this separation of knowledge from its use has resulted in the field of knowledge-based systems being stuck at too low a level of abstraction. The rule-level, frame-level, or logic-level view often obscures the real nature of information processing in complex knowledge-based reasoning. For example, the fact that MYCIN [Shortliffe 76] engages in some form of classification problem solving, while R1/XCON's design strategy [McDermott 82] is one of linear subtasking, is not readily visible at the level of the rule representation, or in the forward- or backward-chaining control strategies used to make inferences.

For several years one of the authors (Chandrasekaran) has been proposing an alternative level of abstraction for knowledge-based system building [Chandrasekaran 83a, Chandrasekaran 86, Chandrasekaran 87]. This is called the *generic task* level of analysis and representation.

This book presents some of our attempts to understand how design can be understood as a collection of generic tasks. A similar analysis was done earlier for diagnostic reasoning [Chandrasekaran 83b, Josephson 87]. Generic tasks are to the rule-level or frame-level analysis what high-level programming language constructs are to assembly language level programming. The generic tasks are an attempt to capture the knowledge and inference phenomena at a more perspicuous level.

Before describing what characterizes a generic task, let us understand the source of the problems implicit in the current generation of languages.

It is intuitive to believe that there are types of knowledge and control regimes that are common to diagnostic reasoning in different domains. Similarly, there should be common structures and regimes for design as an activity. However, one would expect that the structures and control regimes for diagnostic reasoning and design problem solving will be different.

However, when one looks at the formalisms (or equivalently the languages) that are commonly used in expert system design, the knowledge representation and control regimes do not capture these distinctions. For example, in diagnostic reasoning, one might wish to speak in generic terms such as malfunction hierarchies, rule-out strategies, setting up a differential, etc. For design, the generic terms might be device/component hierarchies, design plans, ordering of subtasks, etc. Ideally one would like to represent diagnostic knowledge in a domain by using the *vocabulary* that is appropriate for the task. But the languages in which the expert systems have been implemented have sought uniformity across tasks, and thus have had to lose perspicuity of representation at the task level.

Another source of confusion is that languages such as EMYCIN and OPS5 are computation-universal, i.e., they are equivalent to Turing Machines. That means that any computer program can be written in these languages, more or less naturally. The fact that these languages have been used to construct problem solving systems does not by itself say anything about how appropriate they are, and how natural they are for expressing the knowledge and control structures.

The control regimes that these languages come with (in rule-based systems they are typically variants of forward or backward chaining) do not explicitly indicate the real control structure of the system at the task level. For example, as mentioned earlier, the fact that R1/XCON [McDermott 82] performs a linear sequence of subtasks is not explicitly encoded. The system designer has "encrypted" this control in the pattern-matching control of OPS5. The fact that the low level language is used to implement a higher level control structure results in the so-called knowledge-base actually containing quite a few rules that are really programming devices for expressing this higher level control. That is, the much talked-about separation of knowledge from inference is not really true in practice for complex problems.

These comments need not be restricted to the rule-based framework. One could represent knowledge as sentences in a logical calculus and use logical inference mechanisms to solve problems. Or one could represent it as a frame hierarchy with procedural attachments in the slots. In the former, the control issues would deal with choice of predicates and clauses, and in the latter, they would, for example, be at the level of which links to pursue for inheritance. None of these have any direct connection with the control issues natural to the task.

While on one hand task-level control phenomena are lost in the details of the lower-level implementation, on the other hand, artifacts of the low level representation acquire a theoretical status. For example, rule-based approaches often concern themselves with conflict resolution strategies. Since many rules may be relevant at any one time, a choice will have to be made. That entire analysis takes place at the rule level. Consequently, the solutions offered for conflict resolution have a syntactic feel to them. That is, they emphasize features that do not have a clear conceptual connection to the task-level control issues.

If the knowledge is viewed at the appropriate level, one can often see the existence of *organizations* of knowledge that only allow the selection of a small, highly relevant body of knowledge without any need for conflict resolution at all. Of course, these organizational constructs could be "programmed" in the rule language ("metarules" are meant to do this in rule-based systems). However, rules and their control are often given status as theory-level phenomena in expert systems -- as opposed to implementation-level phenomena, which they often are. Consequently, knowledge acquisition is often directed towards strategies for conflict resolution, whereas they ought to be directed to issues of *knowledge organization*.

The roots of the problem are two-fold. First, separating knowledge from the processes which use it often gives a false sense of generality. In fact the form in which knowledge is needed for design is different from that of diagnosis, so then the designer has to undertake a complex programming effort in order to make this translation. Second, seeking uniform control mechanisms comes at a cost: the architectures that support this uniformity do so by suppressing the distinctions in control and inference between different kinds of tasks.

1.3.1 Characterization of a Generic Task

Each generic task is characterized by information about the following [Bylander 87]:

1. The type of *problem* (the type of input and the type of output). What is the function of the generic task? What is the generic task good for?

2. The *representation* of knowledge. How should knowledge be organized and structured to accomplish the function of the generic task? In particular, what are the types of *concepts* that are involved in the generic task? What concepts are the input and output about? How is knowledge organized in terms of concepts?

The *inference strategy* (process, problem solving, control regime). What inference strategy can be applied to the knowledge to accomplish the function of the generic task? How does the inference strategy operate on concepts?

The phrase "generic task" is somewhat misleading. What we really mean is *an elementary generic combination of a problem, representation, and inference strategy*. The power of this proposal is that if a problem matches the function of a generic task, then the generic task provides a knowledge representation and an inference strategy that can be used to solve the problem.

As mentioned, this book explores some of the generic tasks that can be used to construct a class of design problem solving systems. A number of generic tasks for other problems, such as diagnostic reasoning, have been explored at the Ohio State University Laboratory for AI Research. One of the building blocks of diagnostic reasoning, *hierarchical classification* [Gomez 81], is used in the following section to exemplify the idea of generic tasks.

1.3.2 Hierarchical Classification

Problem: Given a description of a situation, determine what categories or hypotheses apply to the situation.

Representation: The hypotheses are organized as a classification hierarchy in which the children of a node represent subhypotheses of the parent. There must be knowledge for calculating the degree of certainty of each hypothesis.

Important Concepts: Hypotheses.

Inference Strategy: The establish-refine strategy specifies that when a hypothesis is confirmed or likely (the establish part), its subhypothesis should be considered (the refine part). Additional knowledge may specify how refinement is performed, e.g. to consider common hypotheses before rarer ones. If a hypothesis is rejected or ruled-out, then its subhypotheses are also ruled-out.

Examples: Diagnosis can often be done by hierarchical classification. In planning, it is often useful to classify a situation as a certain type, which then might suggest an appropriate plan. The diagnostic portion of MYCIN [Shortliffe 76] can be thought of as classifying a patient description into an infectious agent hierarchy. PROSPECTOR [Duda 79] can be viewed as classifying a geological description into a type of formation. Hierarchical classification is similar to the refinement part of Clancey's heuristic classification [Clancey 84].

1.3.3 High Level Languages Based On Generic Tasks

A "shell" can be associated with each of the generic tasks, since each of them is characterized by types of knowledge and their organization, and a family of inference processes. A toolset [Chandrasekaran 86] can be built consisting of a number of such shells. A Toolset can be used on a problem in such a way that each shell is used to express knowledge for a part of the overall problem, where each subproblem matches the structure of a generic task.

For example, a language called CSRL [Bylander 86] has been built to capture the knowledge for hierarchical classification. Later chapters of this book describe a language called DSPL that captures some of the knowledge for design. These languages facilitate not only knowledge acquisition, but also control of problem solving, since each language makes available the set of appropriate inference strategies. For example, the establish-refine control and its variants for classification are directly available in CSRL, and do not need to be implemented in the control strategy of a lower level language such as OPS5.

We discuss elsewhere how knowledge acquisition [Bylander 87] and explanation of problem-solving [Chandrasekaran 88b] are facilitated by the generic task approach. Thus the general argument is that understanding a complex task such as design in terms of generic strategies that are useful in solving them, characterizing the knowledge and inference requirements of these strategies, and supporting them with high level tools is a good idea that pays off in clarity of system architecture, control of problem solving, knowledge acquisition and generation of explanation.

Precisely how the generic task (GT) idea is applied to create expert systems is best done by means of a concrete example. This book is about the application of this idea to the problem of design. In the next chapter we analyze design problem solving from the viewpoint of generic tasks and in the chapters following it show how the GT view is applied to a version of the design problem that we have called *routine design*. We also describe a tool called DSPL that is useful in the construction of such design systems.

Before we proceed to a discussion of the GT approach for design, we need to round out our discussion of the GT approach itself by outlining some critiques of the GT view and our responses to them, and relate the approach to some other recent work that has similarities to the GT view. These are the tasks for the remaining sections of this chapter.

1.3.4 Critiques of the GT Approach

Is there a downside to the building of knowledge-based systems by using task-specific architectures such as the GT tools? We have often heard the following concerns raised in the research community about the GT approach:

1. Does the GT approach result in multiple representations of the same knowledge?
2. Does the GT view result in difficulties in debugging due to a procedural representation of all knowledge? Does the GT view give up ease of explanation because of non-declarativeness of control knowledge?
3. Is there a loss of flexibility of control due to the need for predetermined control transfers between GT problem solvers?

The following discussion can be understood better after at least Chapter 3 is read, because some of our answers presume some idea of how a GT-based knowledge-based system actually gets built. However, since the book is really about design, this chapter is a natural place to deal with these questions.

1.3.4.1 Multiple Representations of Same Knowledge

The concern can be stated by using the following example. The knowledge about the structure and function of a device may be used both for diagnosis and design. If we have diagnosis-specific architectures and design-specific architectures, then the structure-function knowledge will need to be represented twice.

The answer to this question is that the above could happen if the tasks are not chosen at the right level of abstraction. The way we have been developing the GT view and the architectures to support it, however, we do not use the term "task" to refer to tasks at the application level, such as diagnosis or design. We regard the latter tasks as compound tasks, consisting of more primitive generic tasks, or more precisely, *information processing strategies* that can be used in the service of problems such as diagnosis and design. This can be done if the tasks such as diagnosis or design can be decomposed in the given domain into subtasks for which these strategies are appropriate. Diagnosis for example can often be accomplished by an appropriate combination of subtasks of hierarchical classification, concept-matching and data abstraction. The GT approach has identified generic information processing strategies for each of the subtasks if knowledge in appropriate forms is available.

Coming to the example of the structure-function knowledge about devices and their use both in diagnosis and design, we have identified in [Sembugamoorthy 86] a knowledge representation and a set of inference strategies for representing and reasoning about structure and function of devices. The strategies, for example, can produce malfunction hierarchies for device diagnosis or can simulate the behavior of the device under various assumptions about the structure of the device. Note that the information thus produced by these strategies from the structure-function representation is really knowledge that can be used for diagnosis or design. Thus the structure-function representation and the associated strategies constitute another generic task family that can be a common substrate for diagnosis and design, rather than be necessarily buried in the design or diagnostic architectures. Note, however, that diagnostic knowledge (e.g., the malfunction hierarchy) or design knowledge (e.g., proposals for components, given some functions that the device has to accomplish) that can be derived from the structure-function knowledge is already in a form that is particular to the task of diagnosis or design. The following schematic shows what a proper architecture based on the GT-view will look like.

In fact a number of recent dissertations show how the above relationships between structure-function knowledge, diagnosis and design can be concretely realized. [Sticklen 87] shows how diagnostic knowledge can be derived from the functional representations, and [Goel 88] shows how structure-function knowledge can be used to modify previous designs to achieve new functionalities.

The above figure also shows how the GT view need not result in a proliferation of task-specific architectures. By building particular types of diagnostic or design architectures as higher-level layers over more generic architectures, each layer takes knowledge of greater generality and builds knowledge of more specific use, while retaining the advantages of the generic-task view.

1.3.4.2 Procedural vs Declarative Knowledge Representations

Our implementation of the GT view is, as described in Chapter 3, in the paradigm of a cooperating community of specialists. Thus when the AIR-CYL system to be described is solving a particular air cylinder design problem, an appropriate way of describing it is as follows: ".. The AIR-CYL specialist makes some design decisions and passes control to the PISTON specialist .." But this procedural view is appropriate only at the level of the user of the expert system built using the GT

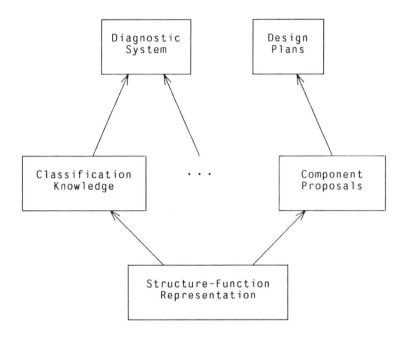

Figure 1: Relation Between Knowledge at Structure-Function, Generic Task, and Higher Order Task Levels

approach. Each of the specialists is built by describing domain knowledge *declaratively* in DSPL and by *explicitly* setting control behavior by choosing the messages to send to the other parts of the system. The DSPL code of a design expert system built using DSPL is a declarative version of the domain knowledge of design plans and their interconnections. It is also a declarative encoding of problem solving and control knowledge, since part of writing DSPL code is to choose the *message* primitives by means of which the different specialists communicate with each other regarding what actions to take. The set of message primitives is domain-independent, but specific to the generic task, and is the language for the specification of problem solving and control strategies. The DSPL interpreter combines the domain knowledge and the message primitives into an interacting collection of problem solvers. The explanation generation mechanisms described in [Chandrasekaran 88b], e.g., examine the declaratively stated domain knowledge in DSPL and the also messages to compose strategic explanations. Thus at the representation level the GT

13

approach enables a declarative representation of domain and control knowledge, even though not in a form that has been popularized in the rule-based paradigm.

In any case it should be noted that the specialist type implementation is just that, an implementation. The GT approach can just as easily be implemented in the rule paradigm, with metarules enforcing the GT strategies. Thus the power of the GT view is not in how it is implemented, but by guiding the knowledge system builder by explicitly offering generic strategies and supporting their use by providing knowledge and control primitives appropriate for each strategy.

1.3.4.3 Flexibility of Control

The GT approach as it presently stands calls for the decision about how to achieve subtask, i.e., what generic task problem solver to invoke, to be made at system building time, rather than at run-time. Further, some of the strategies are made part of other strategies in a predetermined manner. For example, DSPL includes a form of *matching* to choose between plans. Since matching is a generic task, it would contribute to clarity and flexibility if the matcher were to be separated rather than made a permanent part of DSPL.

This is precisely the direction in which GT research is moving. A forthcoming dissertation [Punch 89] investigates run-time integration of generic tasks by explicitly reasoning about the overall task to be solved, the current state of problem solving, and the appropriateness of available strategies. Other investigators in the Laboratory at Ohio State are examining flexible lower level architectures such as SOAR [Newell 87] for implementing the GT approach: the hope is that the resulting use of the strategies will be more fine-grained than the current GT modules, and that subtasks and generic strategies can be matched by reasoning at run time.

1.3.5 Task-specific Architectures

In the 1970's, when Mycin pioneered the use of rule-based architectures for building diagnostic systems, Gomez and Chandrasekaran [Gomez 81] proposed that classification was a basic task that needed to be performed for many diagnostic tasks. This view resulted in the development of the MDX system in 1979 [Mittal 80a, Chandrasekaran 83b], which organized the diagnostic knowledge into a classification hierarchy. The MDX system also demonstrated the usefulness of decomposing diagnosis into a hierarchical classification module and a data-interpretation module, each with its own knowledge structure and inference patterns.

This view resulted in the generic task methodology, which was originally reported on 1981 as a taxonomy of types of problem solving. In later papers, we developed a more thorough critique of the level of abstraction that the rule-, frame- and logic-level languages implied, called for architectures which support the knowledge and inference requirements directly at the task-level, and proposed the generic task methodology as a way of achieving this goal.

In the last several years, interest in task-level analyses has grown, and a number of investigators have made proposals in this direction. Here we would like to describe briefly three such proposals and their points of contact and differences with the generic task approach.

Clancey's work on diagnosis shares the intuition that task-specificity in analysis brings a lot of leverage to system construction and explanation. He proposes that what the diagnostic component of Mycin was really doing was displaying a behavior called *heuristic classification*, which was mapping data into categories. Heuristic classification, in turn, could be decomposed into three subtasks: *data abstraction*, *heuristic match*, and *refinement*.

He has also proposed that diagnostic strategies in the heuristic classification are a collection of micro-tasks organized in a certain order, and that these micro-tasks can be achieved by appropriate metarules in a rule-based system. Finally, he has proposed a language called HERACLES that directly supports the specification of knowledge needed for the micro-tasks, invoking them and combining them to create the heuristic classification behavior for diagnosis.

The spirit of this work, including its concern with task-level analysis as a source of power, and its identification of classification as a major information processing task in problem solving, is so compatible with our own work that elucidating the differences requires concentrating on the more subtle aspects of the underlying views of the architecture of intelligent problem solving.

MDX uses some problem-solving components which together are precisely identical to the functionality of heuristic classification. The components combined are Hierarchical Classification (which performs the *refine* component of heuristic classification), Hypothesis Matching[1] (which performs the *heuristic match* component),

---

[1] or Structured Matching as we have been labeling it recently to emphasize the hierarchical abstraction structures involved in it.

and Knowledge-Directed Information Passing (which performs *data abstraction*).

Notice that by identifying these components as separate types of information processing activities, and specifying their knowledge and inference requirements, it makes it possible for each of them to be used in behaviors other than heuristic classification. For example, Knowledge-Directed Information Passing and Matching are components in other problem-solving activities such as planning and design.

John McDermott and his co-workers have also begun to emphasize task-specific analysis. We can use their SALT tool [Marcus 86] as a prototypical example of their work. SALT is intended as a knowledge acquisition tool for design expert systems that use a *propose-and-refine* strategy. The design method involves constructing an initial design by proposing values for design parameters, identifying constraint violations as design proceeds, and using prestored correction strategies for changing the values of those design parameters which are responsible for constraint violations.

We will not attempt a detailed comparison between SALT and the generic task of *design by plan selection and refinement* that underlies the work to be described in this book. However, broadly speaking, the latter is aimed at a somewhat more complex class of problems, and the DSPL language to be described includes constructs for representing constraint-violation detection and failure handling knowledge. Failure handling could include changing the values of certain parameters. SALT appears to solve the class of constraint-satisfaction-type design problems more thoroughly, i.e., it has a much more exhaustive set of constructs for acquiring information about how to solve constraint failures and correcting them by making changes to parameters.

In more general terms there is a difference in grain size between strategies such as *propose-and-refine* and *design by plan selection and refinement*. The latter makes stronger commitments to a more complex knowledge organization and contains the former strategies, among others, in some form for specific local design actions. That is, the DSPL approach takes the view that the actual problem-solving activity for initial proposing and subsequent refinement of parameters will in general require a more organized knowledge structure and a more complex set of strategies to focus the activity.

Even though all design can in principle be viewed as a propose-and-refine activity (just like all design can be viewed as a rule-based problem solving activity), using it as the exclusive problem solving strategy for increasingly complex design problems would amount to foregoing the advantages of even higher level design strategies that can give more organizational leverage.

The work by Wielinga et al [Wielinga 86] has some points of similarity with the philosophy behind the generic task view. They propose a conceptual analysis of the task as the starting point rather than be concerned with implementation languages, such as rules, which have dominated the discussion in the area of expert systems. In this we clearly agree. They suggest that this stage of the analysis of the problem should be clearly independent of *how* the problem should be solved. The reader will note that the information processing analysis of design problem solving to be discussed in Chapter 2 precisely follows this view. The analysis of what design is is kept separate from a discussion of the methods that can be used to achieve the task.

It might appear that this idea is in conflict with the generic task notion that knowledge and inference should go together. This paradox is resolved by noting that the word "task" has somewhat different meanings in "task analysis" and "generic tasks". The generic tasks provide a *method* for accomplishing each of the subtasks into which a task such as design or diagnosis can be decomposed. As methods they have both a *what* component, i.e, their input-output specification, and a *how* component, i.e., the inference types that they can use. In this book we will show that design is a task which can sometimes be performed successfully by the generic strategy called *design by plan selection and refinement.*

In the following chapter we present an analysis of design from the viewpoint of generic tasks.

2 A Framework for Design Problem Solving

2.1 What is the Design Problem?

In this chapter we look at design as an information processing task: i.e., we specify what kinds of input and output characterize the task abstractly. This can then form the basis for investigating what kinds of knowledge and inference processes can help solve what parts of the task. We will avoid talking in terms of particular methods of representation of knowledge, say, rules or frames, but concentrate instead on *what* needs to be represented, and what types of inferences are needed. Once the nature of the subtasks in design becomes clear, then the question of how to implement them can be undertaken. One of our criticisms of the expert systems area has been that implementation-level phenomena have been allowed to interfere with an analysis of task-level phenomena. We would like to keep them clearly separate.

Design is a very complex activity and covers a wide variety of phenomena: planning a day's errands, theory construction in science, and composing a fugue are all design activities. In order to give some focus and use some shared intuitions, let us restrict the scope in this discussion to the design of artifacts that satisfy some goals.

A designer is charged with specifying how to make an artifact which satisfies or delivers some goals. For each design task, the availability of a (possibly large and generally only implicitly specified) set of *primitive components* can be assumed. The domain also specifies a repertoire of *primitive relations* or connections possible between components. An electronics engineer, for example, may assume the availability of transistors, capacitors, and other components of various types when he is designing a waveform generator. Examples of primitive relations in that domain are serial and parallel connections between components. Similarly, an architect might assume the availability of building materials. If the architect has to design an unusual brick as part of his architectural specification, at least he can assume the availability of clay, and so on.

Of course design can also be recursive: if a certain component that was assumed to be available is in fact not available, the design of that can be undertaken at the next round, even though the domain for the component may be rather different than the original domain, as in the example of building and brick design. If the component design is not successful, the original design may be discarded and the task undertaken again.

The design task can then be specified as follows:
- *Complete specification of a set of 'primitive' components and their relations so as to meet a set of constraints.*

Some of the constraints will refer to the functions or goals of the artifact, some may pertain to the parameters of the artifact (e.g., 'total weight to be less than 1 ton'), yet others may provide constraints on the design process itself, and finally constraints may apply to the process of making, testing or assembling the artifact (manufacturing constraints). Often the goals may not be stated explicitly or in sufficient detail at the start of the design process. In hard design problems, the world of primitive objects may be very open-ended. In spite of all such caveats, the above working definition is a good starting point for our discussion.

This definition also captures the *domain-independent* character of design as a generic activity at some level of abstraction. Planning, programming and mechanical design all share the above definition to a significant degree. Note that the knowledge needed and many of the detailed mechanisms will of course be domain-specific. For example, mechanical design may call for significant amounts of spatial reasoning, while the electrical domain may only involve topological reasoning. But the nature of the design problem as a whole has many commonalities at the level of the above definition, and as we shall see, at the level of many of the subprocesses.

This definition is not meant to imply the existence of one method for all design. The main message of our work is that design actually consists of a large number of distinct processes that work together, each contributing some information needed during the design process. In fact the apparent difference in the design process in different domains and different designers can be explained by the dominance of some of these subprocesses over the others due to differences in the knowledge available.

The above definition gives a clue to why the design problem can be *hard* for AI and often also for people. In realistic domains the size of the set of primitive objects is quite large and the set is not initially made explicit. The design problem is formally a search problem in a very large space for objects that satisfy multiple constraints. Only a vanishingly small number of objects in this space constitute even "satisfying", not to speak of optimal, solutions. What is needed to make design practical are powerful strategies that radically shrink the search space.

2.2 What Kind of Space to Search?

The idea of search in a state space goes back to the early days of AI, and Newell [Newell 80] has formalized the *Problem Space Hypothesis* essentially stating that all goal-directed behavior takes place in some problem space.

Before search can take place, the problem space needs to be defined. But design problem-solving does not have a unique problem space. Different kinds of problem spaces can be visualized, each appropriate for some kinds of domain knowledge and not others. For search in a problem space to be operationally definable there must be problem states, operators which transform one problem state into a set of successors, and some ordering knowledge that helps to choose between alternatives. For search to be practical, generation of successors and choice among alternatives should not themselves be complex problem-solving activities. The last condition means that domain knowledge should be directly available which can be applied to generate successors and choose among them.

Let us consider the *transformation* approach to design [Barstow 84, Balzer 81] as a concrete illustration of these issues. We can consider the set of specifications to be the initial state, and a fully designed artifact to be the end state. Operators transform parts of the specifications into alternative design commitments that will realize them. So an intermediate state will consist of design commitments which realize some of the specifications along with remaining specifications. The process of design can be thought of as searching for a series of design commitments that result in a goal state.

While this is formally satisfactory, knowledge may not be available in all domains for successor generation and alternative selection. In the programming domain to which this idea has been applied, there seem to be several examples where knowledge of this form is in fact available. However, this problem space is not of general applicability -- note that no single problem space is. In some domains, the constraints as stated may not be factorizable in this way, and there may be significant interactions between the designs that are chosen to meet parts of the constraints.

There is also no guarantee that the design process can always correspond to incremental choices. Large subsystems may be designed first and only then can design proceed within subsystems. Thus the actual design process in that domain may not correspond to navigation in this transformation problem space. Knowledge may be directly available which cuts a swath across the space, so that several constraints together are realized by a precompiled design that is recognized as applicable. Finally, in many domains, the problem is reformulated by a decomposition so that a number of disjoint local spaces, each corresponding to a subproblem are created. We will discuss decomposition and design plans shortly in greater detail.

The point of the discussion is this. Which problem space is used depends on the forms in which domain knowledge for representation and control are available. Using an inappropriate problem space will result in artificial heuristic functions being used which do not capture the real structure of domain knowledge.

We propose that in design problem-solving *a variety of types of knowledge* can be identified, each of which helps solve a portion of the design problem in a computationally efficient way. Expertise consists of an accumulation of a repertoire of such knowledge. However, unlike the current view in the expert systems area, this expertise is not viewed as collections of pieces of knowledge, to be used by a uniform inference technique. Instead, knowledge comes in various *generic forms*, each structured in characteristic ways and using their own appropriate inference methods. Each type of knowledge can produce some information that may be needed or useful during design, or can generate a part of the design solution. Conversely, each type of knowledge *requires* information of certain types to be available before it can be useful.

Thus the picture that we would like to give of design problem solving is that of a cooperative activity between multiple types of problem solvers, each solving a subproblem using knowledge and inference of specific types, and communicating with other computational modules or problem solvers for information that is needed for it to perform its task, or delivering information that they need for their tasks. Thus an analysis of design as problem-solving consists of identifying these subprocesses, their information processing responsibilities, and the knowledge and inference needed to deliver these functions. We call this kind of analysis an *information processing analysis* of design. This is the task of the next section.

2.3 Information Processing Analysis of Design

The style of analysis will be to identify subtasks in design, and characterize what kinds of information or solution they are responsible for providing. Some of these subtasks can be performed in a number of different ways: an AI solution is only one way. For example, during design, it will be necessary to find if a certain design requirement is met. A traditional computational algorithm may be able to do that in some cases, e.g., finding out if stress in a member is less than a certain amount may be done by invoking a finite element analysis algorithm. Sometimes this information may require an AI-type solution, involving an exploration of some space in a qualitative way, e.g., by doing a qualitative simulation of the artifact. In what follows we will only describe issues associated with AI-type solutions for these subtasks, but the larger possibility needs to be kept in mind in the actual design of knowledge-based systems for design.

During the discussion we will try to relate the framework to a number of previous and current approaches to design. But the literature on design is vast. Even within AI, work on design has proliferated over the last decade. We do not intend to be exhaustive in our coverage. Our intent is to point to some of the other work as a way to illuminate the discussion.

We will describe a number of subprocesses or subtasks in design and describe the role they play. The design process can be usefully separated into those processes that play a role in the "generate" part and those that help in the "test" part. We subdivide our discussion into two groups of processes: those that are responsible for proposing or making design commitments of some sort, and those that serve an "auxiliary" role, i.e., generate information needed for the proposers, and help test the proposed design.

2.3.1 Processes That Propose Design Choices

(1): Decomposition. This is a very common part of design activity. We will use this process as an example of information processing analysis, and describe it in terms of all the features that such an analysis calls for: types of knowledge, information needed, and the inference processes that operate on this form of knowledge.

Knowledge of the form D --> {D1, D2, ... Dn}, where D is a given design problem, and Di's are "smaller" subproblems (i.e., associated with smaller search spaces than

D) is often available in many domains. In some domains, there may be a number of alternate decompositions available, and so choices (and possible backtracking) will need to be made in the space of possible decompositions. Repeated applications of the decomposition knowledge produce *design hierarchies*. In well-trodden domains, effective decompositions are known and little search at that level needs to be conducted as part of routine design activity. For example, in automobile design, the overall decomposition has remained largely invariant over several decades.

Dependable decomposition knowledge is extremely effective in controlling search since the total search space is now significantly smaller. This power arises from the fact that known decompositions represent a previously compiled solution to a part of the design problem, and thus at run-time the design problem solver can avoid this search.

Information Needed: The decomposition process needs two kinds of additional information for it to be effective.
- How the goals or constraints on D get translated into constraints on the subproblems D1, ... Dn.
- How to glue the designs for D1, D2, ... Dn into a design for D.

Information of the above types may be given as part of the decomposition knowledge or can be obtained by accessing another processor which can produce that information. We will shortly refer to a method called *constraint posting* [Stefik 81] that has been proposed for generating constraints on subproblems. How to glue the designs for subproblems may require additional problem-solving, such as simulating D1 and D2, and finding out exactly where and how the gluing can occur. The CRITTER system [Kelly 82], for example, provides a specialized simulation facility that helps both in generating constraints for the subproblems and in gluing the solutions together.

Inference Process: There are two sets of inference processes, one dealing with which sets of decompositions to choose, and the other concerned with the order in which the subproblems within a given decomposition ought to be attacked. (Remember that a decomposition merely converts a design problem into a set of presumably "smaller" problems, which still need to be solved for the decomposition to be successful.)

For the first problem, in the general case, the decomposition will produce an AND/OR node, i.e., will produce decompositions some of which are alternatives and others all of which need to be solved. Finding the appropriate decomposition may involve searching in a space represented as an AND/OR graph. But as a rule such searches are expensive. Routine design problems should not require extensive searches in the decomposition space. To avoid the search problem while using domain knowledge about decomposition, learned heuristics may be used, or interaction between human experts and machine processing can be arranged so that the machine proposes alternative decompositions, and the human chooses the most plausible ones. Precisely this sort of shared labor is used in the VEXED system [Mitchell 85] during problem decomposition.

The problem of the order in which to attack the problems in the decomposition list, when combined with the problem of searching in the decomposition space, can make the total search very complex. In general, the investigation of the subproblems in a given decomposition will not be reusable if that decomposition is unsuccessful. This explains the extreme difficulty of the design problem in the general case. However, in most cases of routine design, the decomposition knowledge leads to a design hierarchy as mentioned. The default control process for investigating within a given design hierarchy is then *top-down*. While the control is top down, the actual sequence in which design problems are solved may occur in any combination of top-down or bottom-up.

For example, in designing an electronic device, a component at the tip level of the design hierarchy may be the most limiting component and many other components and subsystems can only be designed after that is chosen. The actual design process in this case will appear to have a strong bottom up flavor. Control first shifts to the bottom-level component, and the constraints that this component design places on the design of other components are passed up.

A related issue is one of whether the control should be depth-first or breadth first. Again, this is very much a function of the domain. The specification language for control behavior in this process should be expressive enough for a variety of control possibilities along these lines.

Decomposition is an ubiquitous strategy in AI work in design. McDermott's DESI/NASL system [McDermott 77] uses this extensively. Freeman and Newell [Freeman 71] discuss various decomposition criteria, including functional and structural. The transformational design work of Barstow and Mostow uses decomposition in a degenerate form: the constraint set is such that subsets of it correspond to different design problems, and so can be separately expanded.

(2): Design Plans. Another pervasive form of design knowledge, representing a precompiled partial design solution, is a *design plan*. A design plan specifies a sequence of design actions to take for producing a piece of abstract or concrete design. In abstract design, choices are made which need to be further "expanded" into concrete details at the level of primitive objects.

These design plans are indexed in a number of ways. Two possibilities are to index by design goals (*for achieving <goal>, use <plan>*), or by components (*for designing <part>, use <plan>*). Since plans may have steps that point to other plans, design plans can include decomposition knowledge. From the viewpoint of complexity reduction, the contribution that plans make is as an encoding of previous successful exploration of a problem space. They are produced by abstracting from the experience of an individual expert or a design community in solving particular design problems.

Each goal or component may have a small number of alternative plans attached to them, with perhaps some additional knowledge that helps in choosing among them. A number of control issues arise about abandoning a plan and backing up appropriately, or *modifying* a plan when a failure is encountered.

The inference process that is applicable can be characterized as *instantiate and expand*. That is, the plan's steps specify some of the design parameters, and also specify calls to other design plans. Choosing an abstract plan and making commitments that are specific to the problem at hand is the instantiation process, and calling other plans for specifying details to portions is the expansion part.

A number of additional pieces of information may be needed or generated as this expansion process is undertaken. Information about dependencies between parts of the plan may need to be produced at runtime (e.g., discovering that certain parameters of a piston would need to be chosen before that of the rod), and some optimizations may be discovered at run time (e.g., the same base that was used to attach component A can also be used to attach component B).

For example, NOAH [Sacerdoti 75] can be understood as a system that instantiates and expands design plans. In NOAH, corresponding to each goal of the artifact under design, there is a stored procedure which can be interpreted as a design plan. These plans can call other procedures/plans until a hierarchy of procedures is created. NOAH concentrates its problem-solving on recognizing ordering relations and redundancies between the components of the plan.

The idea of design plans has been used successfully, in the domains of algorithm design and programming, by Rich [Rich 81] and Johnson [Johnson 85]. The notion that *plans* constitute a very basic knowledge structure has been with us from the late 1950's when this idea was discussed extensively by Miller, Galanter and Pribram [Miller 60]. Schank and Abelson [Schank 77] have also discussed the use of plans as a basic unit of knowledge. The Molgen work of Friedland [Friedland 79] uses design plans as a basic construct. More recently Mittal's PRIDE system [Mittal 86] has used them for design knowledge representation.

(3): Design by Critiquing and Modifying Almost Correct Designs. A variation on the design plan idea is that the designer has a storehouse of actual successful designs indexed by the goals and constraints that they were designed to satisfy. Sussman [Sussman 73] has proposed that a design strategy is to choose an already-completed design that satisfies constraints closest to the ones that apply to the current problem, and modify this design for the current constraints. This process needs information of the following kinds.

- *Matching*: How to choose the design that is "closest" to the current problem? Some notion of prioritizing over goals or differences, in the sense of *means-ends analysis*, may be needed if this information cannot be generated by a compiled matching structure. In some cases, some analogical reasoning capabilities may be appropriate by which to recognize "similar" problems.

- *Critiquing*: Why does the retrieved design fail to be a solution to the current problem? This analysis is at the heart of learning from failure, and sophisticated problem-solving may be needed to analyze the failure. The capability of critiquing a design is of more general applicability than for this particular design process.

- *Modifying*: How to modify the design so as to meet the current goals? In many cases, this information may be available in a compiled form, but in general, this also requires sophisticated problem-solving.

The processes of critiquing and modifying have more general applicability than as parts of this particular design process. We discuss criticism as one of the auxiliary processes later in this section. Design modification, however, is a useful process in the "generate" part of design, so we discuss some of the issues related to it here.

Modification as a subprocess takes as its input information about failure of a candidate design and then changes the design. Depending upon the sophistication about failure analysis and other forms of knowledge available, a number of problem-solving processes are applicable:

- A form of means-ends reasoning, where the differences are "reduced" in order of most to least significant.

- A kind of hill-climbing method of design modification, where parameters are changed, direction of improvement noted, and additional changes are made in the direction of maximal increment in some measure of overall performance. This form can even constitute the only method of design in some domains: heuristically assign initial values to the parameters, and change them in a hill-climbing fashion until a maximum is reached, and deliver that as the design. This is especially applicable where the design problem is viewed as a parameter choice problem for a predetermined structure. The system called DOMINIC [Dixon 84] engages in this form of design problem-solving.

- Dependencies can be explicitly kept track of, in such a way that when a failure occurs, the dependency structure directly points to *where* a change ought to be made. *Dependency-directed backtracking* was proposed by Sussman [Stallman 77] as one approach to this problem. Mittal [Mittal 86] uses a variation on dependency tracking for modification of designs on failure.

- What to do under different kinds of failures may be available as explicit domain knowledge in routine design problems. This information can be attached to the design plans. The work to be described in later chapters uses this highly compiled form.

(4): Design by Constraint Processes. For some design problems a process of simultaneous constraint satisfaction by constraint propagation can be employed. In order for this to work computationally effectively, it is best if the structure of the artifact is known and design consists of selecting parameters for the components. Constraints can be propagated in such a way that the component parameters are chosen to incrementally converge on a set that satisfies all the constraints. Mackworth [Mackworth 77] provides a good discussion of several techniques for this. This is an instance of what is called, in optimization theories, *relaxation procedures*[2]. Human problem solvers aren't particularly good at this form of information processing without pencil and paper. The incremental convergence process can be treated as a form of problem space exploration, so we are including it in this discussion.

Constraint satisfaction processes can be viewed as applying design modification repeatedly and incrementally. Thus many of the comments we made earlier regarding

---

[2]Unfortunately, this use of the term relaxation interferes with another use of it in design, viz., *relaxing the constraints* so that a hard design problem may be converted into a relatively easier one.

design modification are applicable. In particular, some of the constraint propagation techniques can be viewed as versions of *hill-climbing* methods in search. Variations such as dependency-based changes to parameters can be adopted during each modification cycle. More complex processes such as *constraint-posting* can be used where additional constraints are generated as a result of choices made for earlier parameter choices. These constraints are used for remaining parameter choices.

Configuration problems are an interesting and well-known class of problems (made famous by the R1/XCON system [McDermott 80]). Some versions of them can be decomposed into subproblems whose solutions can be neatly glued back together. In fact, R1's problem-solving is done as a linear series of subtasks. However, in the general case, these problems often have no clear decomposition into subproblems, because of extensive interactions between various parts of the design. On the other hand, many configuration problems have the tractable feature that most of the components of the device are already fixed, and only their connections and a few additional components to mediate the connections need to be chosen. This makes iterative techniques applicable by making it likely that one can converge on the solution. Constraint satisfaction methods are often applicable to configuration problems.

Marcus and McDermott [Marcus 85] discuss a strategy called *propose and revise*, where commitments are made for some parts of the design, which generates additional constraints, and if later parts in the design problem cannot be solved, earlier commitments are revised. Frayman and Mittal [Frayman 87] discuss the configuration task abstractly.

CAUTION! Formally *all* design can be thought of as constraint satisfaction, and one might be tempted to propose global constraint satisfaction as a universal solution for design. The problem is that these methods are a fairly expensive way to search the space. For example, *propose and revise* can end up searching the entire space in difficult problem spaces, and hill-climbing methods can get stuck at local optima. Hence these methods are not a universally applicable for practical design. Other methods of complexity reduction such as problem decomposition are still very important in the general case. They can create subproblems with sufficiently small problem spaces in which constraint satisfaction methods can work without excessive search.

Human problem solvers need computational assistance in executing constraint satisfaction approaches. This is because the methods are computationally intensive and place quite a burden on short term memory. As long as no attempts are made to use them as universal design methods, they can be effective computational techniques for portions of the design problem.

2.3.2 Auxiliary Processes

The subprocesses in design that we have considered so far are:
- decomposition, design plan instantiation and expansion, modification of an almost correct design, constraint satisfaction.

These contribute to design by proposing some design commitments. Along the way, we have referred to some other processes which serve the former by providing information that they need. Let us discuss them briefly here.

(1): Goal/Constraint Generation for Subproblems. Given a decomposition D --> {D1, D2, ... Dn}, one will need to know how the goals/constraints of D are translated into goals/constraints for the subproblems. In many domains, this information is part of the decomposition knowledge. But if it is not available, additional problem-solving is called for. The literature on constraint-posting that we referred to earlier proposes methods that might be applicable in some cases.

VEXED [Mitchell 85] provides an example of constraint generation for subproblems given a particular problem decomposition. In this domain the subproblems can have a serial connection relation. For example, D may be implemented by two modules D1 and D2 connected in series. A constraint propagation scheme (implemented in a system called CRITTER [Kelly 84]) takes the input to D and produces the constraints on D1's output and D2's input. Design of D1 and D2 can then proceed.

(2): Recomposition. We alluded to recomposition during our discussion of decomposition. The issue is how to glue the solutions of the subproblems back into a solution for the original problem. Integrating them may require simulating the subdesigns to find how they interact. Alternatively, other methods of problem-solving may be called for. Scientific theory building involves assembling portions of theories into larger coherent theories, and needs powerful interaction analyses. RED [Josephson 87] proposes an especially powerful strategy for composing explanatory theories.

(3): Design Verification. This is part of the "test" component of the design activity. That is, whether a candidate design delivers the functions and satisfies any other relevant constraints. In most cases, it can be done by straightforward compiled computational methods (e.g., "add weights of components and check that it is less than x") or by invoking possibly complex mathematical methods, such as a finite element analysis, that do not involve problem-solving. In some cases, additional problem-solving may be needed to complete the verification. For instance, qualitative simulation of a piece of machinery to decide if any of its parts will be in the path of another part may be needed for verifying a proposed design.

(4): Design Criticism. At any stage in design, any failure calls for analyzing the candidate design for the reasons for failure. This form of criticism played a major role in the method of design by retrieving an almost correct design. In most routine design, fairly straightforward methods will suffice for criticism, but in general this calls for potentially complex problem-solving. *Design modification* uses the results of criticism.

2.4 Implications of Above Analysis

The analysis of the design process in terms of subprocesses with well-defined information processing responsibilities has helped us to identify the types of knowledge and inference needed. This in turn directly suggests a *functional architecture* for design with these subprocesses as *building blocks*. It also suggests a principled way in which to define the human-machine interaction in design. Firstly, whenever knowledge and control can be explicitly stated for one of the modules or building blocks, that module can be built directly, by using a knowledge and control representation that is appropriate to that task. Secondly, if knowledge for a module is not explicitly available, the human can be part of the loop for providing information that that module would have been responsible for.

For example, failure analysis and common sense reasoning involving space and time are difficult problem-solving tasks. These tasks may be needed for the performance of design modification and design verification, respectively. The human/machine division of responsibility may be done in such a way that the machine turns to the user for the performance of these tasks. As these tasks are better understood, they can be incrementally brought into the machine side of the human/machine division of labor.

Another kind of human machine interaction is possible in this framework. Note that each subprocess is characterized both by specific types of knowledge and by inference and related control problems. One way in which a module can interact with a domain expert is by proposing available knowledge and letting the human make the control choices by using knowledge that has not been made explicit in the problem-solving theory. As a practical matter, this can be an effective way of using the module as a knowledge source, even without a complete theory of problem-solving using that knowledge. The VEXED system in fact works in this mode. It proposes possible decompositions, and the user is asked to choose the one he or she would like to pursue. Similarly, when a design system's choice of design plans fails, it may turn to the user for choosing alternative plans.

Let us elaborate on the functional architecture for design that results from this analysis. Because each subprocess uses characteristic types of knowledge and inference, a "mini-shell" can be associated with it and knowledge and inference can be directly encoded using that shell. Since each of the tasks has a clear information processing responsibility, the modules can communicate with each other in terms of the information that defines the input and outputs of these modules. Thus the modularity that results is a task-level modularity.

In the rest of this book, we provide the details of the functional architecture for one type of design, a form of *routine design* that we have termed Class 3 Design. This way of analyzing design and identifying architectures out of which design problem solvers can be built is what is novel about the point of view of the book. In the next section we proceed to a description of our informal classification of design problems.

2.5 Classes of Design

The above analysis of design subprocesses can be used to provide an *informal* classification of design problems. Many of the processes in the "test" part of design, such as design verification by qualitative simulation, can be arbitrarily complex, but they are not particularly specific to design. The design process simply calls upon these other problem-solving skills. On the other hand, many of the processes in the "generate" portion are quite specific to design as a problem solving process, so our classification is based largely on the subprocesses in the "generate" part of design.

Each of the processes:
- decomposition, design plan instantiation and expansion, modification of similar designs, constraint satisfaction,

performs some aspect of design, using information either directly available or supplied by auxiliary problem-solving or other computational processes. Each of them comes with a set of control problems that can be more or less complex, and needs knowledge in certain forms.

This classification of design was initially proposed in [Brown 83]. The framework suggests that design by decomposition, i.e., breaking problems into subproblems, by plan synthesis where necessary, and by plan selection where possible, are the core processes in knowledge-based design. That is, it gives importance to the first two processes in the above list as the major engines of complexity reduction in design. The classification is largely based on the difficulty of these subtasks or processes, in particular on the completeness of knowledge, the ready availability of the needed auxiliary information and the difficulty of the control issues.

2.5.1 Class 1 Design

This is open-ended "creative" design. Goals are ill-specified, and there is no storehouse of effective decompositions, nor any design plans for subproblems. Even when decomposition knowledge is available, most of the effort is in searching for potentially useful problem decompositions. For each potential subproblem, further work has to be done in evaluating if a design plan can be constructed. Since the design problem is not routine, considerable problem-solving for many of the auxiliary processes will need to be performed.

The average designer in industry will rarely, if ever, do Class 1 design, as we consider this to lead to a major invention or completely new products. It will often lead to the formation of a new company, division, or major marketing effort. This then is extremely innovative behavior, and we suspect that very little design activity is in this class.

2.5.2 Class 2 Design

Class 2 design is characterized by the existence of powerful problem decompositions. However, design plans for some of the component problems may be in need of *de novo* construction or substantial modification. Design of a new automobile, for example, does not involve new discoveries about decomposition: the structure of the automobile has been fixed for quite a long time. On the other hand, several of the components in it constantly undergo major technological changes, and routine methods of design for some of them may no longer be applicable.

Complexity of failure analysis will also take a problem away from being routine. Even if design plans are available, if the problem solver has to engage in very complex problem-solving procedures in order to decide how to recover from failure, the advantage of routine design is reduced. In short, whenever substantial modifications of design plans for components are called for, or when synthesis in the design plan space is especially complicated, we have a Class 2 problem.

2.5.3 Class 3 Design

This is relatively routine design, where effective problem decompositions are known, compiled design plans for the component problems are known, and actions to deal with failure of design solutions are also explicitly known. There is very little complex auxiliary problem-solving needed. In spite of all this simplicity, the design task itself is not trivial, as plan selection is necessary and complex backtracking can still take place. The design task is still too complex for simple algorithmic solutions or table look up.

Class 3 problems are routine design problems, but they still require knowledge-based problem-solving. The following chapters of this book deal with an approach to building knowledge-based systems for routine design problems of this type. The processes described here can work in conjunction with auxiliary problem solvers of various types, but the theory for them is not developed further in this book. The examples used all assume that the information to be provided by the auxiliary design processes, e.g., design criticism, verification, and subproblem constraint generation, are all available in a compiled manner.

2.5.4 A Class 3 Product

In a large number of industries, products are tailored to the installation site while retaining the same structure and general properties. For example, an Air-Cylinder intended for accurate and reliable backward and forward movement of some component will need to be redesigned for every new customer in order to take into account the particular space into which it must fit or the intended operating temperatures and pressures. This is a design task, but a relatively unrewarding one, as the designer knows at each stage of the design what the options are and in which order to select them. Note that that doesn't mean that the designer knows the complete sequence of steps in time (i.e., the trace) in advance, as the designer has to be in the problem-solving situation before each decision can be made. There are just too many combinations of requirements and design situations to allow an algorithm to be written to do the job.

As this tends to be unrewarding work for humans and as this type of non-trivial problem appears to be possible to do by computer there is strong economic justification for us to attack this problem.

2.5.5 Class 3 Complexity

The complexity of the Class 3 design task is due not only to the variety of combinations of requirements, but also to the numerous components and sub-components, each of which must be specified to satisfy the initial requirements, their immediate consequences, the consequences of other design decisions, as well as the constraints of various kinds that a component of this kind will have.

While Class 3 design can be complex overall, at each stage the design alternatives are not as open-ended as they might be for Class 2 or 1, thus requiring no planning during the design. In addition, all of the design goals and requirements are fully specified, subcomponents and functions already known, and knowledge sources already identified. For other classes of design this need not be the case. Consequently, Class 3 design is an excellent place to start in an attempt to fully understand the complete spectrum of design activity.

Note that we are not merely interested in producing an expert system that produces a trace which is the same as or similar to a designer's, nor are we solely interested in arriving at the same design -- although both are amongst our goals. We are concerned with producing an expert system that embodies a theory of Class 3 design and demonstrates the theory's viability.

2.5.6 Imprecision of the Classification

The classification that we have described is a useful way to get a bearing on the complexity of the design task, but it is *not* meant to be formal or rigorous. Neither is the term *routine design*. The approach described in this book is intended to provide a starting point for capturing some of the central phenomena in routine design, but it is not intended to be a complete account of routine design.

3 Expert System Architecture for Class 3 Design

3.1 The Structure of a Class 3 Design Problem-Solver

In this book we will consider a Class 3 design problem-solver to consist of a hierarchical collection of design specialists, where the upper levels of the hierarchy are specialists in the more general aspects of the component. The lower levels deal with more specific subsystems or components. They all access a **design data-base** mediated by an intelligent data-base assistant [Mittal 80b] [Chandrasekaran 80][3].

We will first describe the structure of the design agents[4] (i.e., the types of knowledge), and then their problem-solving action and the phases of their interaction. To attempt to make the following sections clearer we will illustrate the descriptions of the types of knowledge in the system with examples taken from the AIR-CYL system. These examples are written in the Design Structures and Plans Language (DSPL). Some details of the examples will not be explained here, as it is premature. They will be more fully explained in Chapter 5.

3.2 Design Agents

3.2.1 Specialists

A **Specialist** is a design agent that will attempt to design a section of the component. Specialists will control the problem-solving. The specialists chosen for a design system, their responsibilities, and their hierarchical organization will reflect the mechanical designer's underlying conceptual structure of the problem domain. Exactly what each specialist's responsibilities are depends on where in the hierarchy it is placed. Higher specialists have more general responsibilities. The top-most specialist is responsible for the whole design. A specialist lower down in the hierarchy

---

[3]The "intelligence" in this data-base is demonstrated by constraint checking applied to values inserted into the data-base. No complex inferences appear to be necessary at this stage of the development of the theory. However, the full power of an intelligent data-base could be used, if it were found to be necessary, without affecting the rest of the theory, as the data-base is decoupled from the design problem solving being done.

[4]By the term "Agent" we mean any active module of the problem-solving knowledge.

will be making more detailed decisions.

Each specialist has the ability to make design decisions about the part, parts or function in which it specializes. Those decisions are made in the context of previous design decisions made by other specialists. A specialist can do its piece of design by itself, using local design knowledge captured in tasks (see below), or can utilize the services of other specialists below it in the hierarchy (Figure 2). We refer to this cooperative design activity of the specialists as **Design Refinement**, as each specialist contributes more details to the design, thus refining it.

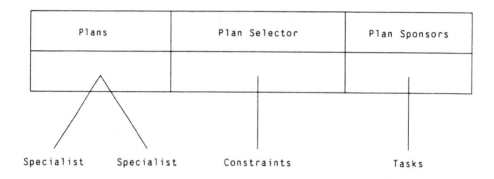

Figure 2: A Specialist

Every specialist also has some local design knowledge expressed in the form of constraints (see below). Constraints will be used to decide on the suitability of incoming requirements and data, and on the ultimate success of the specialist itself (i.e., the constraints capture those major things that must be true of the specialist's design before it can be considered to be successfully completed). Other constraints, embedded in the specialist's plans (see below), are used to check the correctness of intermediate design decisions and the compatibility of subproblem solutions. Still more constraints are present in the design data-base as general consistency checks.

3.2.1.1 Specialist Structure

Figure 3 shows the DSPL for the specialist for the Air-cylinder Head from the AIR-CYL system.

A specialist knows which specialists it can use and which specialist can use it -- that is, it knows its place in the conceptual hierarchy in a restricted way. It may have constraints to test for the applicability of the specialist on entry, and others that will test conditions on exit. If the success of a specialist depends on some essential design

```
(SPECIALIST
 (NAME Head)
 (USED-BY AirCylinder)
 (USES None)
 (COMMENT "for the Head of the AC")
 (DESIGN-PLANS HeadDP1)
 (DESIGN-PLAN-SELECTOR  HeadDPSelector)
 (ROUGH-DESIGN-PLANS HeadRDP1)
 (ROUGH-DESIGN-PLAN-SELECTOR  HeadRDPSelector)
 (INITIAL-CONSTRAINTS None)
 (FINAL-CONSTRAINTS None)
)
```

Figure 3: Specialist "Head"

values then those can be tested by the entry constraints (INITIAL). The exit constraints (FINAL) may exist to check that everything is correct before reporting success. As these are generally useful all design agents can have exit or entry constraints.

There are two sets of plans, separated into those for design and those for rough design. These phases will be discussed later. Associated with each specialist's set of plans is a "Plan Selector" that will select the next plan. Plan selection is an important part of the design process and will be discussed in Section 3.8. The specialist also has Failure Handlers and a Redesigner. These will be discussed more fully in the next chapter. For now it is sufficient to know that Failure Handlers will be used if something is discovered to be wrong with the design while attempting to follow a plan.

3.2.2 Plans

Each specialist has a collection of plans that may be selected from depending on the situation. A specialist will follow a plan in order to achieve that part of the design for which it is responsible. A **Plan** consists of a sequence of calls to Specialists or Tasks (see below), possibly with interspersed constraints. It represents one method for designing the section of the component represented by the specialist. The specialists called will refine the design independently, tasks produce further values themselves, constraints will check on the integrity of the decisions made, while the whole plan gives the specific sequence in which the agents may be invoked.

As each plan is considered to be the product of past planning, refined by experience, one should not expect many failures to occur. However, as not all combinations of values have been handled before or anticipated, it is possible for plan failures to occur due to intra-plan and extra-plan constraint violations. This is discussed in section 4.5.4.

3.2.2.1 Plan Structure

Figure 4 shows the DSPL for a design plan used by the Head specialist in the AIR-CYL system.

```
(PLAN
 (NAME HeadDP1)
 (TYPE Design)
 (USED-BY Head)
 (SPONSOR HeadDP1Sponsor)
 (QUALITIES Cheap)
 (INITIAL-CONSTRAINTS None)
 (FINAL-CONSTRAINTS None)
 (BODY HeadTubeSeat
     MountingHoles
     Bearings
     SealAndWiper
     AirCavity
     AirInlet
     TieRodHoles
     (REPORT-ON Head)
) )
```

Figure 4: Plan "HeadDP1"

Constraints can test to see if a plan considers itself to be applicable to the situation it has been selected for, and, on completion to see if its design goal has been achieved. The main part of the plan is an ordered list of plan items to be executed by the specialist. A plan item can be a Constraint, a Task, or a Specialist. Specialists can be asked to do design or to do rough design. A plan item can also be two specialists used in parallel. Clearly this requires the operation of the specialists to proceed independently and for there to be no interactions or dependencies between them. This would realistically occur with a single human designer when another person is asked to design some part because of their greater expertise or because it would save time. It also corresponds to a situation where the designer shares his or her attention between two parts of the design in a disorganized way (i.e., the pattern of attention has no significance in terms of problem-solving).

The order of items in a plan will depend partly on the designer's preference, partly from experiences that show that that order minimizes problems, but mainly from the underlying dependency structure in the design. A design agent can be said to be dependent on another agent if it uses a value that was decided by that agent. Dependency can be defined for specialists, tasks, and steps.

The items in a plan must follow dependency ordering, as one can't use something before it has been decided. If the plan is <p ; q ; r>[5] then there are several different dependency patterns for which this is a reasonable plan, including no dependencies. For example, notice that if "q is depending on p" is the only dependency, then the plan could equally well be <p ; r ; q> or <r ; p ; q>. Dependency knowledge is discussed in Section 3.2.7.

3.2.3 Tasks

A **Task**[6] is a design agent that is expressed as a sequence of steps with interspersed constraints. It is responsible for handling the design of one logically, structurally, or functionally coherent section of the component; for example a seat for a seal, or a hole for a bolt. A task has little affect on control during problem-solving, but acts as an organizational unit for the specialist to which it belongs, as it will coordinate the action of some group of basic design decisions (i.e., steps).

3.2.3.1 Task Structure

Figure 5 shows the DSPL for a task in the AIR-CYL system. The task is responsible for the Air Cavity in the Head.

```
(TASK
 (NAME AirCavity)
 (USED-BY HeadDP1)
 (BODY
  AirCavityDepth
  AirCavityID
  AirCavityOD
  CheckAirCavity
))
```

Figure 5: Task "AirCavity"

The main part of a task is a sequence of task items that are to be executed by the task. A task item is a step or a constraint (Figure 6). To handle errors that arise

---

[5] We will use angled brackets to delimit plans, and semicolons to separate plan items.

[6] Unfortunately the word "task" here does not have much in common with the word in the phrase "generic tasks" introduced in Chapter 1. It is a technical term that describes a part of design plan.

during execution of the task items, the task has Failure Handlers and a Redesigner. The task can also have a collection of suggestions that are also used during failure recovery.

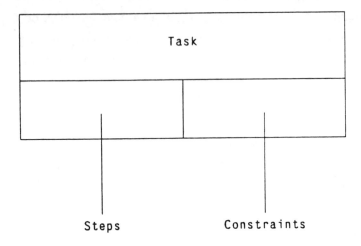

Figure 6: A Task

For the moment a task can be considered as a kind of degenerate specialist. The task items form a fixed plan with limited capabilities. As a consequence all of the plan selection knowledge is missing from a task. A task is only concerned with communicating with steps below. A task with its steps represents design knowledge local to a specialist. It also is restricted to contributing to the design something with only a few attributes to be decided, whereas a specialist is more properly associated with a larger collection of attributes, such as a part. These differences will be further discussed later.

3.2.4 Steps

We consider a **Step** to be a design agent that can make a single design decision. For example, one step would decide on the material for some subcomponent, while another would decide on its thickness. A step provides a value for an attribute of the design. The step is the primitive design agent. It is the only agent that directly makes decisions about attributes. A step cannot use other agents directly, but can use values from the current state of the design. It makes no decisions about control during the problem-solving.

3.2.4.1 Step Structure

Figure 7 shows the DSPL for a step in the AIR-CYL system. The step designs the Internal Diameter of the Air Cavity in the Head.

```
(STEP
 (NAME AirCavityID)
 (USED-BY AirCavity)
 (ATTRIBUTE-NAME  HeadAirCavityID)
 (REDESIGNER AirCavityIDRedesigner)
 (FAILURE-SUGGESTIONS
  (SUGGEST (DECREASE RodDiameter))
  (SUGGEST (DECREASE HeadBearingThickness))
  (SUGGEST (CHANGE HeadMaterial
            TO DECREASE MinThickness))
 )
 (COMMENT "Find air cavity internal diameter")
 (BODY
  (KNOWN
   BearingThickness (KB-FETCH
            'Head 'HeadBearingThickness)
   RodDiameter (KB-FETCH 'Rod 'RodDiameter)
   HeadMaterial (KB-FETCH 'Head 'HeadMaterial)
   MinThickness (KB-FETCH
            HeadMaterial 'MinThickness)
  )
  (DECISIONS
   MaxRodRadius       (VALUE+ (HALF RodDiameter))
   MaxBearingThickness (VALUE+ BearingThickness)
   AirCavityRadius    (+ MinThickness
                (+ MaxRodRadius
                 MaxBearingThickness))
   AirCavityID  (DOUBLE AirCavityRadius)
   REPLY       (TEST-CONSTRAINT ACID)
   REPLY       (KB-STORE
            'Head 'HeadAirCavityID AirCavityID)
  )
))
```

Figure 7: Step "AirCavityID"

The step obtains values from the design data-base, and has step actions that may be constraints, calculations, or simple choices. Most steps store a value in the design data-base as their last action. A step has Failure Handlers, a Redesigner, and a collection of suggestions, all of which are used to handle failure during design.

3.2.5 Constraints

A **Constraint** is a piece of knowledge that will be used to test for a particular relationship between an attribute and some constant, or between two or more attributes. For example, a constraint might check that a hole for a bolt is not too small to be machinable given the material being used. Constraints can occur before or after any action in the design knowledge, as any action can produce a value and any value may have bounds. Other constraints act in a more global way and are attached to the design data-base. Constraints will be discussed further when we address failure handling.

Ideally, constraints would be embedded in the design knowledge wherever there is a test that can be made to confirm that the last calculation or decision will not lead to failure. In practice, many such opportunities will be missed by the human designer as they are not immediately obvious, or because they are too time consuming. In an expert system those constraints can be included to provide better performance. Once the design knowledge is expressed explicitly by the designer more constraints will be noticed and included.

3.2.5.1 Constraint Structure

Figure 8 shows the DSPL for a constraint in the AIR-CYL system. It checks the size of the Internal Diameter of the Head's Air Cavity relative to the size of the Internal Diameter of the Tube.

A constraint has a unique identifier or message that distinguishes it on failure from other possible failures in the design knowledge. This is necessary as there can be more than one constraint in a step, and, if the step fails due to a failing constraint, it should be possible for the supervising task to respond differently depending on which constraint failed. The constraint collects necessary information about the state of the design and tests some relationship. As the test may not be directly between two attributes there may be some small amount of calculation involved. The constraint also includes some suggestions that are used when the constraint fails.

```
(CONSTRAINT
 (NAME ACID)
 (USED-BY AirCavityID)
 (FAILURE-MESSAGE
  "Air cavity ID too large relative to tube ID")
 (FAILURE-SUGGESTIONS
  (SUGGEST (INCREASE TubeID
            BY FAILURE-AMOUNT))
  (SUGGEST (DECREASE HeadAirCavityID
            BY FAILURE-AMOUNT) )
 )
 (BODY
  (KNOWN
   TubeID  (KB-FETCH 'Tube 'TubeID)
  )
  (TEST
   ( AirCavityID
     <
   (TubeID - (DOUBLE MinThickness))
   )
)))
```

Figure 8: Constraint "ACID"

3.2.6 Types of Design Knowledge

We will assume that there will be only one specialist associated with any one part or subpart. Any variations in the design will arise from the plan selected. How might variations occur? Different plans can use different tasks. Tasks used in a plan are considered to be local to the specialist and part of the specialist's skills. Different plans could access different specialists. This would mean that the object being designed would have some allowed major variation that warrants another specialist being used.

Class 3 design specialists may require plans to refer to the same specialists below -- e.g., <S2 ; S3 ; S4> and <S3 ; S2 ; S4> -- or may not -- e.g., <S2 ; T1 ; S4> and <S3 ; T1 ; S5>. The latter would occur when the problem S1 actually decomposes more than one way; here into subproblems S2 and S4, or, alternatively, into subproblems S3 and S5.

Even with a unique decomposition it is still possible to have variation due to tasks in the plans. For example, <S1 ; T1 ; S2> and <S1 ; T2 ; S2> might both design two blocks connected by a bar, but T1 may provide a square bar, while T2 may provide a round bar. Note that this is different from tasks merely varying due to method (i.e., T1 and T2 producing the same design with different values due to two different methods).

3.2.7 Dependency and Dependency Measures

There are two kinds of dependency knowledge that a designer has. One is knowledge about the identities of other agents on which an agent Z depends. The other is which agents are dependent on agent Z. The first kind of knowledge allows the designer to reorder plans and to decide initially how to divide things into specialists, tasks and steps. The second kind of knowledge is the notion of an agent's influence; i.e., possible ramifications of decisions it reaches. Dependency relations are concerned with the first kind of knowledge. Dependency measures are concerned with the second kind. Dependencies may be used in plan selection.

Note that dependencies can cut across the specialist hierarchy. That is, a specialist in one branch can be dependent on the results from a specialist in another branch.

3.2.7.1 Dependency Relations

A designer can have knowledge about agent/agent dependencies. These dependency relations are abstracted from the underlying step/step dependencies and are probably much less accurately represented; except when it is important to the progress of the design, in which case it will have been worked out in detail once and then remembered. One such situation is plan reordering. Plan reordering is the rearrangement of the order of items in a plan, and may be done after plan failure if the dependencies make it possible. Dependency relations are associated with plan items. One reason to reorder would be that there was good evidence that an item in that plan may fail, in which case that item may be advanced in the plan so that that failure can be discovered earlier and less effort wasted.

3.2.7.2 Dependency Measures

A designer has some knowledge about how much effect a local change will have on the surrounding design, and, if it has not been learned from experience, it can be worked out. The "working out" need not be done in a detailed way. It could be done by inspecting the blueprint. The strength of the influence, the dependency measure, is stored as part of the knowledge associated with an agent. Not every agent will necessarily have a measure associated with it. It depends on the need. The measure can be used during plan selection or failure handling.

Designers do not have complete models of dependency, only partial ones. They can build complete but temporary portions of the model when details of dependencies are needed. Results from such an effort may get included in the designer's current model.

3.3 DDB: The Design Data-base

By analyzing the way the designer works and by looking at the content of the drawings used by the designer it is clear that a layered design data-base is required. This allows a designer to make changes with various levels of commitment. The "outer" level can be considered to be the most doubtful. Once more evidence has been accumulated that a value is likely to be correct then it can be collected with related values and made more permanent. Eventually all the current collections of values will be adopted to form a new state of the data-base. By structuring the data-base this way updates, and deletions upon discovering a failure, can be made more efficiently.

3.3.1 Drawings

The drawings (blueprints) used by the human designer have a section for the drawing proper and a section to record "revisions" -- i.e., what alterations were made to the previous drawing to arrive at the current one. A revision is a collection of "alterations", where each alteration is a new value for single attribute. Each alteration in a revision is an Addition, Deletion or a Change, with a Change recording the previous value.

3.3.2 An Hypothesis

An analysis of the protocols suggests that a revision on a drawing corresponds to a task, and that a new drawing corresponds to the completion of a plan[7]. If this were true we would expect the designer to produce as many drawings as there were completed plans; however, it is unlikely that a designer would actually work that way due to the effort involved in producing drawings. We suspect that designers actually produce rough drawings on paper and in their heads until a fairly complete design is

---

[7]Clearly this assumes that our analysis of the structure of knowledge for the Air-cylinder design is correct. Although it will certainly not be correct in all details we have seen nothing so far that suggests that it is wrong, and we have some indications that confirm our analysis.

arrived at, or at least one that would warrant doing a drawing. Notice that this is a "performance" consideration and that it does not necessarily invalidate our hypothesis about the underlying activity.

3.3.3 Alterations and Revisions

A step makes a single alteration, as it is making a decision about the value of a single attribute. An alteration is not considered to be part of a drawing until a task has successfully completed and a revision is made. That is, once the values of a coherent collection of attributes are decided then there is enough evidence to include them as a revision. Note that during a task several alterations may refer to the same attribute, as design may be followed by redesign and other design attempts. The designer keeps track of these alterations in his or her head and in rough form on paper. Only the last one will actually be recorded as part of the design.

A task might have to undo some alterations because that path through the design failed. Similarly revisions may need to be undone as a later element in the plan may fail and cause re-design of that task. Once a plan is completed the collected revisions are used in making a new drawing.

The design data-base (DDB) is structured to reflect the distinctions between the alteration, revision and drawing. Values required by steps, constraints and redesigners are obtained by first looking for them amongst the collected alterations, including revisions yet to be used in an update of the drawing, and then by looking at the drawing, i.e., the design data-base in the implementation.

3.3.4 DDB Constraints

Some knowledge used during design is really not part of the process of design and is not design knowledge per se, but rather knowledge about the type of object being designed. This knowledge is expressible in terms of constraints and manifests itself during design when the designer has just made a decision about the value of an attribute. For example, a constraint could determine if the weight or volume of a subpart was too large for that type of object.

This kind of constraint is triggered by the designer actually or mentally inserting the value into the design -- i.e., making an alteration. The constraint will be "brought to mind" when the drawing is inspected or the attribute's place in a subcomponent is being considered. The constraint's activation will produce an "Ah but ...." response from a designer as this extra piece of knowledge is seen to be relevant and the constraint is tested.

It is hard to define exactly what the difference is between this kind of constraint and constraints explicitly in the design. It is clear that the data-base constraints will often be about more global attributes and their limits, and will be tests that one would not normally include explicitly in one's design procedures. However, if a piece of constraining knowledge is activated in this jack-in-the-box fashion often enough, the designer will tend to remember it and include it in the design activity. This is especially true, of course, if it has often failed. Consequently, as a result of experience, constraints can "migrate" from the knowledge about objects to the knowledge about the design process, and will be explicitly tested in particular places in the design [Brown 86].

3.4 Other Agents

In general, a collection of design specialists will not be sufficient for the design task. Other specialists outside the design specialist hierarchy could provide calculations, such as stress analysis, and other data-base functions such as catalogue lookup. In a more general design system, requests could be made to other types of problem-solvers [Chandrasekaran 83a].

It is perfectly acceptable to consider a human as one of the problem-solvers, as the need for assistance will arise at well-defined points in the design with precise pieces of design to do. The system can subsequently use constraints to check the acceptability of the results provided by the human. The usual image of the designer controlling the invocation of analysis packages and problem solvers is reversed when the system asks the human for assistance.

3.5 The Action of a Class 3 Design Problem-Solver

In this section we will describe the information processing that takes place when doing design problem-solving. Each type of knowledge that has already been described will contribute its action to the overall problem-solving activity. There are four **phases** of design activity.

3.5.1 Requirements Checking

The design activity can be considered to fall into four phases. Initially, the **requirements** are collected from the user and are verified both individually and collectively. For example, it may be reasonable to ask for a component to be made of lead, and for it to weigh less than 5 ounces, but the combination will often be unreasonable. These requirements are available for use by all of the design agents during the design activity. Once it has been established that the system is capable of working with those requirements, a rough-design is attempted.

3.5.2 Rough-design

Rough-design is poorly understood at present, but it serves at least two purposes. First, those values on which much of the rest of the design depends will be decided and checked. If they can't be achieved then there is little point going on with the rest of the design. This also has the effect of pruning the design search space, as once the overall characteristics of the design are established it reduces the number of choices of how to proceed with the rest of the design. Second, as any mutually dependent attributes can prevent a design from progressing (i.e., A depends on B, and B depends on A), rough-design can, as human designers do, pick a value for one of the attributes and use that as if the dependencies didn't exist. This "starter" value can be refined later during the design phase if necessary.

It appears at present that rough-design and design share the same conceptual hierarchical structure. However, that remains to be confirmed. The rough-design hierarchy is in general much shallower than the design hierarchy as more general decisions are being made.

We are proposing that specialists have both design and rough-design plans from which to select. Not all specialists will need both. The knowledge used in rough-design is separate from that used in design. It is entirely feasible that phases could be intermixed during problem-solving, but we have chosen to restrict the rough phase to be completed first, followed by the design phase.

At present, there is no difference in the theory between a task or step in the design phase as opposed to the rough-design phase. However, as rough-design becomes better understood we expect that steps will be able to produce truly "rough" values and that steps and tasks will have simplified failure handling abilities. Specialists too may be able to be simplified.

3.5.3 Design

Once rough-design is completed satisfactorily, the **design** phase can proceed. Design starts with the topmost specialist and works down to the lowest levels of the hierarchy. A specialist S begins by receiving a design request from its parent specialist. It refers to the specification knowledge and data relevant to its further work. A plan is selected using these data and the current state of the design. For example, if one of the requirements is low cost, a plan with that quality can be selected. Plan selection is presented in Section 3.8.

Thus, specialist S fills in some of the design, and calls its subordinate specialists in a plan-directed order with requests for the design of substructures. Parts of a plan may indicate immediately that constraints cannot be satisfied. This is considered as failure. When all of a specialist's plans fail, or when failure can be deduced immediately, the specialist communicates that to its parent.

3.5.4 Redesign and Re-design

If any failures occur during the design process then a **redesign** phase is entered. If the phase succeeds then a return can be made to the design phase. Failures occur when a constraint fails. An agent attempts to handle all failures at the point-of-failure before admitting defeat and passing failure information up to its parent. A step, for example, may be able to examine the failure and then produce another value, in order to satisfy the failing constraint, while still retaining local integrity. By **re-design** we mean the process of asking an agent to attempt design again after previous failing attempt has been recovered from.

Other work on redesign in the literature has concentrated on "functional redesign", that is, "the task of altering the design of an existing, well understood circuit, in order to meet a desired change to its functional specifications" [Mitchell 83]. Here we use **redesign** to mean an attempt by a design agent to change a value to satisfy constraints while keeping as much as possible intact of the previous design.

3.6 Inter-agent Communication

Control and information is communicated from specialist to specialist by passing messages up and down the hierarchy. There is also local communication between a specialist and its tasks, and between a task and its steps. In this way the flow of control is restrained and the system exhibits clear, well-focused problem-solving activity. It remains to be shown whether this form of control is sufficient, but it is based on a belief that Class 3 design systems are "nearly decomposable" [Simon 81] and that "the interactions between subsystems are weak but not negligible". We believe that for Class 3 design the structure is dominantly hierarchical and that interactions are handled by specific strategies.

Information is passed in the form of messages that can, for example, request action, report failure, ask for assistance, and make suggestions. This rich variety of messages is the key to handling subsystem interactions. In addition, one part of the emerging theory of design problem-solving will be the form and content of these messages.

Messages are most informative after failure. The principle is that if an agent below was successful there is no need to know how it achieved that result. However, if failure occurs, every piece of information that can be used to analyze the failure and identify the situation as one which can be recovered from is of use.

3.6.1 Types of Messages

There are the following categories of messages:
- Action Requests
- Reports of Results
- Queries

Action requests are from one agent to another asking it to carry out some kind of action. The actions currently included are Design and Redesign. It may be necessary to include Re-design eventually as the theory grows to include a good account of an agent's use of local memory, but currently it is considered equivalent to a Design request.

Reported results are of Success or of Failure. Success messages flow up the hierarchy from plans, constraints or agents. They contain no information about how success is achieved. Failure messages also flow up the hierarchy from plans, constraints or agents. Each agent that fails will pass up a description of how it failed. We consider every failure to have some kind of description attached to it that uniquely identifies the failure. Any suggestions made by an agent after failure are also included in its failure message.

In the AIR-CYL system each agent that fails will pass up, attached to the message describing its failure, any failure information from below. Any local attempt to recover from failure that itself fails is recorded as an explanation in the agent's failure message.

Queries are currently limited to the question "Are you affected by changes to these attributes?". This is used during redesign when a backup leads to an alteration that might affect attributes already designed prior to the failure. Given a list of attributes an agent will respond with a success message if it is not affected and with failure if it is.

3.7 Design Agent Action

3.7.1 Specialist Action

A specialist is responsible for supervising some portion of the design. Specialists consider design situations and produce courses of action. This includes activating its local design agents (tasks) and calling other specialists. The courses of action lead to changes in the state of the design. A specialist is responsible for any failures that occur during the period it is in control.

The specialist, when activated, will check its initial constraints to see if all the required conditions for applicability have been satisfied. If they are not, it will fail. If they are, it will obtain the plans for that phase (e.g., Redesign) and select a plan (see Section 3.8). The plan is then executed and upon successful completion the final constraints are tested to ensure the specialist's success. Any failures will be processed by failure handlers in order to either "give up" or to attempt failure recovery.

3.7.1.1 Specialist vs. Task

Specialists are responsible for the flow of control during design problem-solving. The selection and execution of plans provide these control decisions. By selecting a plan the specialist is refining the plan that calls it. Specialists therefore are gradually "inserting" plans into other plans in order to "construct" the plan that will produce a successful design. In fact, of course, no actual construction takes place. Figure 9 shows the plan that has been selected by specialist S0 being refined by specialists S1 and S2, while S3 refines the plan in S2.

In contrast, tasks supervise the design activity carried out by steps. This activity is local to the specialist, i.e., a task belongs to a specialist. A task makes no control decisions that affect the overall problem-solving flow. No plan refinement takes place. A task always attempts to carry out the same series of actions. The main role of the task in the system is an organizational one -- that of grouping some related steps and executing them in order. This difference leads to major differences in failure handling

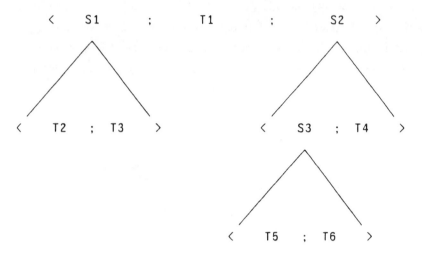

Figure 9: Plan Refinement

between the two types of agents.

We can now see why it is not entirely correct to view a task too literally as a degenerate specialist as suggested earlier. That analogy was made earlier in order to emphasize the major structural differences.

3.7.1.2 Plan Action

A plan is the result of prior decisions about the flow of control in a section of the design. It is a small piece of "frozen" control knowledge. A plan could, for example, specify that an Air Cylinder's Piston and Rod should be designed first in parallel with the Cap, and next the Tube should be designed in parallel with the Bumper.

A plan is followed by testing its applicability conditions (i.e., constraints), by executing each plan item in turn and then checking the exit conditions. A plan item can be a task, a constraint, a design request of a specialist, a rough-design request of a specialist, or an indication that two specialists should be used in parallel. Task execution and constraint testing will be described below.

3.7.2 Task Action

A task activates and monitors the execution of steps. The task will use the steps to design some related set of attributes. It makes no choices. It makes no design decisions itself. For example, it may be responsible for the design of the seat in the head for the tube. It will use steps that design the inside diameter, outside diameter and the depth of the seat. It is also responsible for controlling recovery from failure of those steps over which it has control.

On entry to a task the applicability of the task to the situation is tested. Each task item is executed in turn and, when they have been completed, the constraints which check the task's exit conditions are tested. A task item is a constraint or a step.

3.7.3 Step Action

A step considers a design situation and produces a new situation that has one additional attribute specified. Each step is responsible for one attribute.

On entry to a step, and on exit, constraints can be tested to show the applicability of the step and to see if the exit conditions are satisfied. A step can retrieve values from the design database. These values are used in the decision-making part of the step, which includes all the calculations, simple choices and constraints that are needed to arrive at a single value. If the step is successful the value can be incorporated into the design.

3.7.4 Constraint Action

A constraint tests to see if some design-related value is within some limit. It takes a small portion of the state of the design and returns "acceptable" or "unacceptable". Its task is to prevent the design from becoming unacceptable by recognizing such situations. It takes no other action.

On activation any values that are needed are obtained from the design database. Some values may be immediately available because of the situation in which the constraint is being used. Any calculations which are necessary are made and then the constraint's test is carried out. This is a test to see if a relationship between two values is true. The values can come from calculations or directly from attributes of the current state of the design.

3.8 Plan Selection

Specialists do plan selection in order to affect the flow of control. Different plans will produce different sequences of actions. Plan selection depends on three types of information: the qualities of the plans themselves, the initial demands of the user (requirements) and the situation in which the selection takes place (i.e., the state of the design and its history). We propose a general method of plan selection that responds to all factors.

A designer may have very simple selection criteria, such as "if there is a plan that hasn't been tried yet then try it", or they may be very complex. For example, "if there are some plans that look perfect for the situation then if plan X is amongst them then use it, otherwise pick the one that has been the most reliable in the past, unless it contains the item that failed in the last plan".

The selection process will select one plan from several. Some plans will not be suitable for consideration. Other will, but with various degrees of suitability. Selection then will occur after the individual plans available in the specialist have been evaluated for suitability. Notice that in the example above, the selection is from plans that have already been rated as worth selecting. Consequently, we will divide the whole selection process into evaluation and selection processes, each with its own knowledge.

First we will discuss the process of selection and then the knowledge that can be used during plan selection. An example of the system doing plan selection can be found in Appendix D. The reader is urged to follow that example to obtain a better understanding of the problem-solving involved.

3.8.1 The Selection Process

We propose that plan selection divides into two parts -- first the recommendation of those plans that are candidates for use, and second the selection of a plan from the set of candidates. Each plan has associated with it some information about the qualities of the plan and a Sponsor. It is the job of the plan's sponsor to use its current situation, the qualities of the plan, the user's preferences about qualities, and special case information to make an evaluation of whether this plan is a suitable candidate for selection. It will give an evaluation of the plan's suitability. The Selector has the job of collecting the responses from the sponsors, evaluating them, promoting or relegating if necessary, and selecting one plan for use.

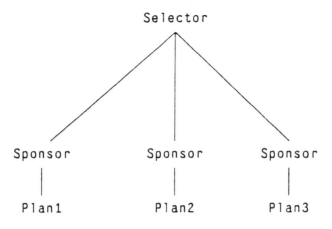

Figure 10: Plan Selection

Sponsors will respond with some scale of suitabilities such as (*perfect, suitable, don't know, not suitable, rule out*). Those ruled out will never get used, unless the Selector has strong reasons to do so. If more than one plan is perfect then the selector will pick one. If none are perfect, then a suitable plan will be used. The selector can do plan quality relaxation, if it is allowed, in order to select from amongst the lower rated plans.

In many situations, there will be few plans, or the suitabilities will always be the same, and consequently this kind of effort during plan selection is not always warranted. In these cases the order can be fixed[8].

3.8.1.1 Sponsors

Figure 11 shows an example in DSPL of a Sponsor.

Each sponsor will be associated with only one plan as it will have knowledge about the applicability of that plan to different situations. Every plan has a sponsor. The output of a sponsor is one of the values from the scale of suitabilities. The inputs to a sponsor are the various sources of knowledge already outlined above. Notice though that some knowledge, for example plan complexity, will only belong in the selector, as it will be used to choose between equally suitable plans.

In a sponsor, evidence needs to be accumulated about the plan's suitability for use. Some information can be used to rule out (or in) a plan quite quickly, while other

---

[8]In the Air Cylinder Design System, AIR-CYL, plans are selected in a fixed order.

```
(SPONSOR
 (NAME   ExampleDPSponsor)
 (USED-BY ExampleSelector)
 (PLAN   ExampleDP3)
 (COMMENT "gives suitability")
 (BODY
  COMMENT "rule out plan if already tried"
  REPLY (IF (ALREADY-TRIED? PLAN) THEN RULE-OUT)
  COMMENT "rule out plan if plan DP1 failed"
  REPLY (IF (ALREADY-TRIED? 'DP1) THEN RULE-OUT)
  COMMENT "use qualities to get suitability"
  Qualities
   (TABLE (DEPENDING-ON
          (RELIABILITY-REQS)(COST-REQS))
      (MATCH
       (IF (Reliable  Cheap)THEN PERFECT)
       (IF (Medium      ? )THEN SUITABLE))
      (OTHERWISE RULE-OUT))
  COMMENT "was Task2 the last failure?"
  Agent   (EQUAL 'Task2 LAST-FAILING-ITEM)
  REPLY
   (TABLE (DEPENDING-ON
          Agent Qualities)
      (MATCH
       (IF ( T    ?    )THEN RULE-OUT)
       (IF ( ?  PERFECT )THEN PERFECT)
       (IF ( ?  SUITABLE )THEN SUITABLE))
      (OTHERWISE DONT-KNOW))
))
```

Figure 11:   Sponsor "ExampleDPSponsor"

pieces of information need to be put together to build a picture of suitability. Consequently, two kinds of expressions of knowledge are needed: one with immediate result and one that accumulates a result.

The immediate form of knowledge is a rule -- for example "if this plan has already been executed during this use of the specialist then it must have failed and should not be considered at this point"; that is, its suitability is "Rule-out". The accumulated form of knowledge will weigh the answers to several "questions", such as "is a COST=Cheap required?", and combine them to produce a suitability value according to some designer-dependent logic[9].

---

[9]This is not meant to imply that predicate calculus is used, but rather that under some circumstances the designer will make "suitable" AND "perfect" produce "perfect", while at others it will produce "suitable".

As there are several different types of knowledge we will propose that evidence is accumulated for each type, e.g., qualities, and then these pieces of evidence are combined to form an overall suitability to represent the plan. This approach has already been demonstrated in the CSRL language [Bylander 83] where diagnostic knowledge is factored into knowledge groups. These groups allow "rule" knowledge or "table" knowledge, corresponding to the two types outlined above.

3.8.1.2 The Selector

Figure 12 shows the DSPL for an example of a Selector.

```
(SELECTOR
 (NAME   ExampleDPSelector)
 (USED-BY Example)
 (TYPE   Design)
 (USES   DPSponsor1 DPSponsor2)
 (COMMENT "gives name of plan")
 (BODY
   COMMENT "if Plan 26 is perfect use it"
   REPLY (IF (MEMBER 'Plan26 PERFECT-PLANS)
        THEN 'Plan26 )
   COMMENT "if there are perfect plans
          use in preferred order"
   REPLY (IF PERFECT-PLANS
        THEN (DESIGNER-PREFERENCE
              PERFECT-PLANS)
        ELSE NO-PLANS-APPLICABLE)
))
```

Figure 12: Selector "ExampleDPSelector"

Each selector will be associated with a specialist and will have a collection of subordinate sponsors. There is one selector for every plan selection situation. A selector can contain specialist dependent information. The input to a selector is the collected outputs from all of the sponsors that report to it, and the state and history of the design. The output consists of the name of the plan and its suitability. So, for example, the information received might be

{ (plan1, perfect) (plan2, suitable)}.

The output from the selector consists of either the name of the plan selected for execution by the specialist, a failure due to there being no plans appropriate, or a failure due to all plans having been tried already.

The exact operation of a selector will depend on the designer's own personal preferences and experiences. The "normal" knowledge would take the plans ranked as "perfect" and choose one. If there are no perfect plans then "suitable" ones will be considered, and so on. Exactly how many of the suitability categories will be considered as acceptable can be made to depend on any appropriate factor, for example, on the position of the specialist in the design hierarchy. A specialist at the lowest extremes can afford to try plans that are less appropriate, as not much effort is being wasted if they fail (i.e., there aren't many agents below). However, at higher levels there are many specialists below and any relaxation of standards could be very costly.

If several plans appear to be equally suitable the designer is most likely to pick the one that has performed the best in the past. That is, the designer has an order of preference. Another approach is to compare some quality (for example COST) and pick those with the best value (COST=Cheap). This can be repeated with other qualities (such as WEIGHT=Light) until one plan remains. Notice that there is no need to prescribe a global ordering for qualities using this method, as orderings will be local to specialists and sensitive to the situation.

Knowledge that can be used during selection includes plan complexity, existing preference, position in the hierarchy, special rules about the use of particular plans (e.g., "if plan A is perfect and it hasn't been used before then use it before any others"), dependencies, and knowledge about past plan failures.

Quality Relaxation is done automatically using this approach to plan selection. Perfect plans will have all the qualities required, while suitable plans will be less ideally matched to the requirements. The combination of how sponsors categorize plans and the degree to which a selector is prepared to compromise on the suitability of the plan provides this relaxation -- i.e., a requirement may be COST=Cheap but a plan with COST=Medium may get selected.

3.8.2 Qualities of Plans

It is not possible to prescribe in advance exactly which of the pieces of knowledge described in the sections below will be used in which cases. What we are arguing here is that these types of knowledge exist, that an expert system builder should be provided with language in which to express them, and that this knowledge must all be available to be used by the plan selection mechanism (i.e., the plan selector and associated sponsors). The theory acknowledges that there are these types of knowledge but does not and cannot describe exactly how each will be used, as it will vary depending on the domain and specialist involved.

Plans can have qualities associated with them. These qualities can refer to some attribute of the plan, some attribute of the design, or some attribute of the object being designed. A similar set of qualities were used by Friedland [Friedland 79] in his MOLGEN system.

- **Precision:** A plan with precise measurements is more expensive in manufacturing terms, and may be harder to design as there is less "slack" in the plan.

- **Convenience:** Convenient plans will have easier calculations, less difficulty with tolerances, fewer elements, fewer other specialists used, and less anticipated trouble.

- **Reliability of design:** Some plans will be associated with reliable products as they capture methods that produce reliability.

- **Reliability of plan:** A designer will know the likelihood of success for a plan. This will have a general component, (i.e., works fairly often), and a context dependent component, (i.e., fails often if Aluminum is the material). A plan that works often will be considered reliable.

- **Cost:** An expensive plan produces an expensive product.

- **Designer's time:** If the plan takes a long time to follow due to many calculations, many steps, many questions of the user, or many catalogue lookups it will be noted as taking a lot of the designer's time.

- **Manufacturer's time:** A note is made if a plan takes a lot of the manufacturer's time.

- **Plan Complexity measures:**
    - Length of Plan: If all else is equal then the designer can be expected to choose the shorter plan.

    - Complexity from Structure: Assuming that it is preferable to select a plan that is in some way "simpler" than another, a "cheap" estimate of complexity is useful. It is possible to obtain some crude measure of the complexity of a plan by using just its surface syntax -- i.e., without detailed knowledge of the structure or action of the components of that plan.

    - Complexity from Dependencies: Another measure takes into account the designer's knowledge of agent-agent dependencies (see previous discussion in Section 3.2.7). Consider two plans, < S1 ; T1 > and < S2 ; T2 >. If the first plan has a measure of 7, and the second a measure of 4, the first plan has more ramifications if adopted, and might therefore be worth avoiding.

- **Manufacturability:** If the processes involved require much skill, unusual machines, special techniques, special materials, and unusual attention to detail then the plan would be classified as difficult to manufacture.
- **Weight:** If the manufactured product falls toward the high end of the range of reasonable weights the plan would be classified as heavy.

3.8.3 Situation Factors

In addition to plan qualities, selection depends on the situation at the time of selection. Information that may be relevant includes:
- The active plans above (in the hierarchy).
- Plan selection information from above.
- Plan selection information from below.
- Which plans were selected already by this specialist and how they failed.
- What has been included in this design (e.g., optional subparts).
- The current state of the design (i.e., values chosen), including those values from rough design.

A specialist in a plan may be passed information on activation. This might include information about the currently active plan, or information about higher plans. For example, if a particular plan quality has not been specified by the user, then a specialist can select a plan according to some preferred quality. In this situation it is good to have specialists below select plans that are in some way compatible -- there is little point selecting a COST=expensive plan inside a COST=cheap plan.

There may be more subtle plan interactions, where from experience it has been discovered that while in plan x selection of plan y is to be preferred. As a specialist has no knowledge of which plans are being executed at higher levels this kind of reasoning can only occur when the specialist above passes some identification of the plan and that identification is known to the lower specialist. We feel that this is less likely than the plan quality case already presented.

Of interest too is the position of the specialist, i.e., the one doing plan selection, in the plan that contains it. If it is towards the end of the plan, then it would be nice not to waste the effort expended so far, and some effort can be made to relax the selection criteria a little and prefer simple and successful plans, rather than insist rigidly on particular qualities. The specialist executing a plan needs to pass down the position in the current plan of the specialists it is invoking. This can be acted on if they choose to do so.

The history of the selection process is important for subsequent selections. Primarily, of course, one does not wish to select the same plan again. Moreover, one does not want to select a plan that is similar to others that have failed. It is possible that the structure or properties of plans that failed could be abstracted out and used to indicate which others to avoid.

The way plans failed is also of use. Not only can there be knowledge such as "If plan A failed then so will plan B", but also more subtle knowledge such as "If plan A failed due to task 1 then plan B will fail", or even "If plan A failed due to xyz being too large then plan B will fail". Knowledge can also be in a positive form so that failure of a plan suggests the selection of another. All of this needs a representation of reasons for a plan's failure to be associated with the plan. The plan item that caused the failure and an abstracted form of its reasons for failure should be available, along with the constraint that precipitated the failure.

Selection may depend on the past choice of values. For example, "If xyz < 0.5 then use plan B". As small variations in the component can be introduced by different tasks, selection can depend on this too -- "If cross section of connecting rod is rectangular then try plan A otherwise try plan B". These kinds of selection rules could actually be expressed as constraints in the plan, so that the use of a plan would be accepted or rejected on entry. However, after continued use this knowledge would migrate so that it could be used during the selection process.

3.8.4 Plan Complexity

The complexity of a plan will be known and associated with the plan. The measure will be in some kind of symbolic scale such as (High, Fairly-high, Medium, Fairly-low, Low). The complexity can be accessed and used during plan selection.

The components of a plan are Tasks, Constraints, or Specialists used for design or rough-design. The complexity of a plan is some function of the complexity of the plan items. If we take the constraint as the major source of problems in the system, then we can assign a rough complexity to each type of agent depending on how many constraints we expect to find in each on average.

```
Constraint    = 1
Task          = 5   (i.e., 5 constraints)
RD Specialist = 20
     (about 4 tasks per RD specialist)
D  Specialist = 25
     (about 5 tasks per D  specialist)
```

It must be stressed that it is not being suggested that numerical measures such as these actually exist, but rather that, with experience, some feeling of the complexity of a particular plan can be formed based on general knowledge about plans, on dependency knowledge, and on the surface structure of that plan. These complexities can be compared. Plan selection can use complexity measures as one of the available sources of knowledge.

3.9 Summary

This chapter presented the types of knowledge necessary for Class 3 design. The structure of Specialists, Plans, Tasks, Steps and Constraints were presented along with the essential relationships between them. The importance of dependencies was also outlined. The structure of the Design Data-Base was presented.

Next we presented Requirements Checking, Rough-Design, Design, and Redesign. The problem-solving action of Specialists, Tasks, Steps and Constraints were explained, along with the message passing that links them. Plan Selection by Selectors and Sponsors was also described.

In the next chapter we discuss how to handle failures that occur during design problem-solving. Failure Handlers and Redesigners are introduced.

4 Failure Handling in Routine Design

4.1 Introduction to Failure-Handling

Our approach to a theory of how failures are handled during design is colored by beliefs that the knowledge available for use in failure handling is restricted, that a social metaphor is applicable, that failure handling is mainly a local rather than global process, and that failure handling processes are domain-driven. We will discuss each of these in turn below.

4.1.1 Restricted Knowledge

Much of the work on failure handling in the literature considers <u>all relevant knowledge</u> to be available at failure time. If one views the problem solver's complete internal model as the "state-of-the-world", then, as one has complete knowledge of the form of the model and one knows that it completely captures the state of the world, it is easy to do any kind of model manipulations that one desires [McDermott 77]. This will lead to the use of unrealistic failure handling mechanisms in AI systems. By structuring the model to reflect its use in problem-solving, model manipulations are constrained. A structured model leads to structured failure handling. We view failure handling as being a *complex structured activity*.

In addition, if all information is globally available there is the problem of finding the relevant information after every failure. If the model is structured in some way, so that at a failure point only part of the whole model is available (i.e., that pertaining to just the most local problem-solving) then the relevant information will be more immediately accessible.

The proposed structure of design problem-solving (i.e., specialists, plans, tasks and steps) provides the context in which to structure failure handling. We will assume that at any point in the structure only the minimum knowledge is available locally about the problem-solving task.

We will restrict the information passed to an agent from above to that which does not provide history but merely makes requests, provides requested information, gives suggestions, and possibly passes constraints. The information received from subagents is restricted to reports of success or failure, and suggestions, with a minimum of information about what took place at the lower levels of the problem-solving structure, except where required by failure reporting.

4.1.2 Social Metaphor

We will continue to use the social metaphor when discussing failure handling -- that is, we can learn about possible behaviors of an agent in the system by considering it to be a person working in a design team organized with the same structure as we are suggesting for Class 3 design[10]. By using this idea, and the minimum-knowledge restriction discussed above, we hope to establish what is essential for failure handling in this kind of design activity.

This metaphor has proven very useful in other work on problem solving in the AI Group at Ohio State University, especially for diagnosis, where efficient knowledge structuring and control strategies can be observed in the medical community [Gomez 81]. We feel that there is much to be gained by applying this strategy to the design domain too.

4.1.3 Local Decisions

We are proposing that all design agents detect their own failure, be able to determine what went wrong (at least superficially), attempt to fix it locally, do so if they can, and report failure only if all attempts fail. Agents that have some control over other agents can use those agents in their attempt to correct the detected problem. The local decisions principle is compatible with the principle of restricted knowledge and the social metaphor described above.

4.1.4 Domain-Driven

Any mechanisms that are adopted in order to handle failures should arise naturally from an analysis of the domain. Different mechanisms will be appropriate in different places. Failure handling will be a complex and varied collection of processes. Different types of failures will lead to different types failure handling.

In general we feel that any wholesale adoption of an AI mechanism will often lead one astray when analyzing a problem-solving situation. For example, use of

---

[10]A discussion of this metaphor can be found in [Chandrasekaran 81] and other papers in that issue.

completely global dependency structures and pure dependency-directed backtracking[11] appears to be inappropriate in this design domain, as it would be unconstrained use of a mechanism in a way that does not reflect the structure of the problem-solving activity or domain knowledge. This should not be interpreted to mean that we consider that belief revision behavior does not occur in humans, rather that analysis of the domain and the task should lead one to it, and that the mechanisms are surface forms of richer problem-solving behavior. In fact, our analysis of failure handling in the mechanical design domain leads us to a backtracking method that does depend on dependencies, and would appear to be dependency-based when viewed from a global point of view.

4.2 Knowledge for Failure Handling

In order to be able to handle failures during design the knowledge structures already presented must be augmented to allow discovery of exactly what failure occurred and to attempt recovery when appropriate.

4.2.1 Failure Handlers

When a failure occurs during the design process it can be due to many causes. For example a piece of design knowledge may be inappropriate for the current situation, or some assisting agent may discover that its part of the design couldn't be done. The design agent needs to inspect the failure reports to see what kind of failure occurred and then prescribe the appropriate action.

A Failure Handler (FH) is the entity that we are presenting as appropriate to do this job. An FH is a piece of knowledge placed in agents at places where a variety of failure messages are expected. Thus specialists, tasks and steps have FHs. For example, a task will have an FH that handles messages from failing subordinate steps. Any design agent can use one of its FHs by passing a failure message to it. That FH will return a success or failure message.

When an FH is presented with a failure message it will use knowledge of situation-action associations. An FH will use the associations to attempt to recognize the failure situation and produce a decision about what action to take next. The action taken may be to report failure, because the failure described by the message cannot be

---

[11]A clear and concise introduction to and bibliography for techniques such as these, referred to as "belief revision", can be found in [Doyle 80].

handled locally; to pass the message to another FH for a more detailed classification of the type of failure; or it may be to recommend an attempt at recovery from the failure.

Failure handlers can match the situations they know about against selected parts of the failure information. For example, one FH may match against a description of the type of failure, while another can match against some part of the history of the failure.

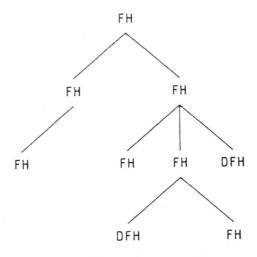

Figure 13: Failure Handlers

Failure Handlers can be considered to be classified along several dimensions. A terminal FH is one which is a tip node of the FH decision tree of which it is a part. FH's can be considered to be design problem dependent or independent. An independent FH (referred to as a System FH in the implementation) is one which would be appropriate for any design problem being tackled, as it is independent of problem (i.e., component) specific failures[12]. For example, the detection of any failing step is done by an independent FH, but the recognition of the failure of a particular constraint is done by a Dependent FH (DFH in Figure 13).

---

[12]In the implementation, the System FHs are already provided, while the person using the design language should supply FHs to take care of the problem dependent failures.

4.2.2 Recovery from Failure

There are two ways of recovering from failure during Class 3 design. The first is by plan selection, and the second by redesigning. Plan selection is the abandonment of the current design approach and the selection of a different and hopefully more successful plan. **Redesigning** is an attempt to use the description of a failure to guide the alteration of an existing piece of the design in order to alleviate the source of the failure. The basic design problem-solving control strategy is first to attempt redesign and then resort to plan selection, so that as much as possible of the results of previous efforts can be saved.

The method of recovery depends on the agent involved. At the step level recovery involves redesign knowledge that is used to attempt to alter the value of the step's attribute. A task has a redesign strategy that will use the steps below in redesign and design modes. For failure during plan execution, recovery will also involve a strategy that can request redesign of the items in the plan.

4.2.2.1 Step Redesigners

As all decisions about design values are made by steps, the most important redesigners are associated with steps. Steps have a designer and a redesigner associated with them. The designer's job is to produce the most reasonable choice of value for an attribute, consistent with the requirements and the current state of the design. The step redesigner's job is to attempt to alter that value just enough to allow all the attached constraints to be satisfied still, while removing the cause of some failure.

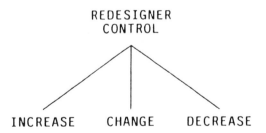

Figure 14: Step Redesigner Structure

The knowledge in a redesigner is in many ways similar to that of the designer part of the step; that is, it fetches known values, does calculations, makes choices, and tests constraints. However, there are important differences.

First, it has sections of knowledge that correspond to whether it is increasing, decreasing or changing the value in question (see Figure 14). As the knowledge required to do these three actions is usually different, especially in the tests or constraints that are applied to see if the alteration is successful, the redesigner is considered to be divided in this way.

Second, for each attribute it is possible to associate with it a minimum adjustment that would be expected to make some difference. If the redesigner gets information that it should change a value by some amount in order to fix some problem, but that amount is less than the minimum adjustment, then the minimum amount is used instead. Typically this would be considered to be the smallest machinable change for that attribute.

A third important difference is that a redesigner will receive information about alterations that will help to fix a problem in the design. Before being used to select the appropriate section of the redesigner this information must be checked for consistency and reduced to a single piece of information that can be used directly. This "suggestion reduction" problem is discussed in Section 4.5.3.2.

The fourth and final difference is that a redesigner needs a value to alter. If the redesigner is being used within a step's execution (i.e., the step's design knowledge failed) then the value will be immediately available, whereas if it is being used as a result of some higher level redesign activity then the value must be obtained from the current state of the design.

4.2.2.2 Other Redesigners

Redesigners other than step redesigners are responsible for controlling the requests for redesign from lower previously executed agents. They do no actual manipulation of values, but will supervise the backtracking inherent in redesign. The task redesigner will be working just with the steps over which it has control and which have already executed. The specialist redesigner will be working with the already executed items in a plan.

There appear to be a limited number of basic strategies used during low level redesign, and a limited number of sources of knowledge. Any one of them can be used in a particular agent for a particular design problem. There is no one redesigner strategy. However, this should not be taken to imply that there is no overall problem-solving strategy to failure handling. The Failure Handler, Redesigner, Plan Selector approach will be outlined below, and in the following sections we will discuss how this approach can use various knowledge sources to produce variations on the basic problem-solving behavior.

4.3 Redesign Problem-solving

4.3.1 How Redesign Occurs

Each kind of agent can have different kinds of reasons for failing. Some have been mentioned above, while others include a task's discovery that a step's failure can't be mended locally, a plan with failing items, and a specialist with all of its plans failing. We would expect most failures ultimately to be caused by a constraint failing; probably, but not necessarily, in a step. That is, failures will be design problem dependent.

Another possibility is that some kind of organizational and content-independent failure occurs. For example, design knowledge is discovered to be incomplete, or missing, or the inter-step dependencies have not been adequately taken into account, leading to the need for some information in a situation where that information has yet to be determined.

Except for cases where redesign is known to be a waste of time or not possible, a Redesigner, if asked by an FH, will attempt to recover from failure by altering a value or values. Redesigners will be given any appropriate "suggestions" taken from the failure messages. These suggestions guide the action of the redesigner.

How a step's redesigner, for example, makes an alteration to a value is clearly very dependent on the current design problem, whereas the way in which a task's redesigner orders its suggestions and requests redesign from its subordinate steps, or how it reacts to their failure during redesign, is based more on problem-independent design knowledge, and is a design problem-solving strategy; that is, a particular kind of control knowledge. We are arguing here that, although particular strategic knowledge will be used by the task during the solution of a particular problem, much of the knowledge used is general to design and not dependent on that particular problem.

4.3.2 Design vs Redesign vs Re-design

As well as being concerned with design we are also concerned with re-design -- i.e., design again. In some cases it may be necessary for an agent to do a re-design instead of a redesign. For example, if during failure handling so much has changed since the last attempt that it makes no sense to even try redesign, then design should be attempted again, i.e., re-design. To decide this locally requires local (in programming terms "static" or "own") memory. Another decision method, and the one we propose, is that a higher agent keeps track of the attributes that have been redesigned and will ask lower agents if they are affected by those changes. This requires agents to have knowledge of the attributes they use in their own decision making. This is the approach that has been implemented.

Compared to design, the step acts quite differently during redesign. Typically redesign is more concerned with testing that a value is not at the extremes of its range, whereas design is more concerned with finding a reasonable and safe central value.

In addition, the suggestions it receives guide the process of deciding a value. Suggestions will be reduced by some kind of "clashing" against each other to produce a single suggestion that summarizes the others. This might include a single value specifying the amount to alter the existing value, or it might include a range of alterations[13]. It may include a direction of change (e.g., increase) or none (e.g., change). Local knowledge is then used to make an alteration and then test it. Different knowledge is required and different tests made depending on the alteration made.

4.4 An Overview of Failure Handling

Before we present the details of failure handling it is appropriate to present a brief summary of the failure handling theory. Failures can occur in step, task, plan or specialist.

When a task or specialist fails their controlling plan's execution is interrupted and a redesigner takes over to attempt recovery. If recovery is achieved the plan continues happily, otherwise the plan fails. A redesigner controls recovery by one of a limited number of backup strategies that use appropriate local knowledge and also the

---

[13]The other possible result is that the suggestions are incompatible and that there is no value.

suggestions received. These suggestions are provided by the failing agent.

When a plan fails, due to a redesigner failing to achieve recovery, the specialist will ask its plan selector to provide another plan. This selection process can use knowledge of the past failures. If no plan can be found to execute then the specialist executing the plans will fail. This leads to another, higher plan attempting redesign.

A task fails when a step over which it has control fails and the subsequent recovery attempt fails. Recovery is done by a task redesigner that uses one of a limited number of strategies to backup from the failing step. This backup is driven by local knowledge and suggestions from the failure. The redesigner will request redesign from any of the task's steps that have already been executed. Which ones get asked is dependent on the suggestions. If the redesigner is successful then the task continues as if nothing had gone wrong.

Steps may fail when something goes wrong during the design, such as a failing constraint. A step will attempt to recover from this failure by using its redesigner. The redesigner will make an alteration guided by the suggestions generated at the point of failure. If it cannot recover then the step will fail. However, if this redesign is successful the step acts as if nothing had gone wrong.

4.5 Design Agent Failure

4.5.1 Constraint Failure

Behind most failures are constraint failures. As already discussed, a constraint will collect information about the state of the design and will test some relationship. Associated with every constraint is a collection of suggestions about how the values involved in the constraint can be changed to make it succeed. In a collection of problem-solving knowledge that is the result of experience these suggestions would be pre-formed, but in general they could be worked out from the structure of the constraint and its context.

A constraint-failure message includes these suggestions and will also indicate the values that caused the failure. This information is formed entirely locally to the constraint and it is up to the agent in which the constraint is embedded to interpret the message and to determine its own actions based on this message and local information.

4.5.1.1 Suggestions

If a constraint tests to see if "x is less than y" then failure may be due to x being too large or y being too small[14]. In which case one suggestion would be to increase y by some amount and the other would be to decrease x by some amount. The amount to be used depends on by how much the constraint fails. For example, if x is 0.9 and y is 0.7 then altering either x or y by slightly more then 0.2 will alter the result of the failing test. Suggestions are always about design attributes.

As well as requesting an increase or decrease it is possible for a change to be suggested. This could arise when it is obvious that some attribute should be changed but the direction is not clear. This is likely to occur in the suggestions from a failing step or task. A change that might occur in a constraint would be a request to change some component's material in order to increase or decrease its minimum thickness. For example, with a harder material the distance between two cuts can be smaller, allowing larger cuts or closer centers.

4.5.2 Failure in a Step

Before investigating Task Failure we should understand the way that steps fail, as steps are the components of a task. A step makes a design decision -- e.g., picks a material or a dimension. It contains two parts: getting needed information, and then using that information to compute and record the decision. The step acts like a "black box" to the rest of the system -- it produces a result or a failure by some method known only to the step. If the step makes choices then only the step knows about them.

Once a failure has occurred at some point in the step, the design action of the step will stop. The failure information is collected together to form a description of the failing situation. This failure message will be passed to the failure handler responsible for failure in that step.

In the case of problem dependent constraints failing a dependent FH (DFH) will be reached. In most cases this will lead to a request for redesigner use. Other possible actions are the production of complaints (e.g., in the implementation a complaint would result in a message being printed to the user), and failure.

---

[14]It may also be due to both being wrong. As we feel that conjunctive failure goals in redesign lead to quite different problem-solving strategies we have left this issue for further research. As the conjunctive case includes the single goal case the latter must be studied first.

The FH gathers together the relevant suggestions for the redesigner (see Figures 15 and 16). If the failing constraint makes no suggestion about the attribute over which the step has control (i.e., what it is trying to design) then the FH will not bother to activate the redesigner and will fail. If there is no redesign knowledge available, or it if is known that redesign is not possible then the redesign fails.

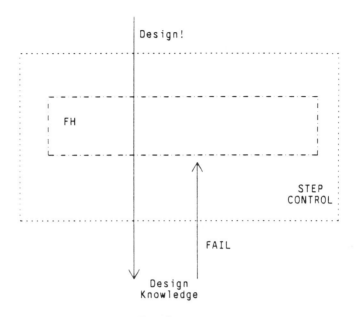

Figure 15: The Step's Failure Handler

The redesign proceeds by altering the current value for the attribute being designed. That is, the suggestion is carried out. Following this are calculations to prepare for constraints being tested, and simple decisions about success or failure. If everything is successful the value will be recorded as part of the design.

If a redesigner fails is it possible to invoke the redesigner again? Under what circumstances should this occur? Certainly the initial suggestion for the redesigner and the suggestion due to the redesigner failing must be compatible. This corresponds to a human designer saying that if I just alter it a little more then it should be OK.

If the redesign succeeds then the step has produced a value for its attribute and it reports success. If the redesigner fails then the step fails. The step's failure message can include the message from the redesigner as part of its explanation of the failure. A failing step can make suggestions about what might be done to correct the problem.

4.5.3 Failure in a Task

We consider every task to have two failure handlers. One is used at the task-item failure level and will take care of failures of steps or inter-step constraints. The other is at a higher level and will take care of all other failures.

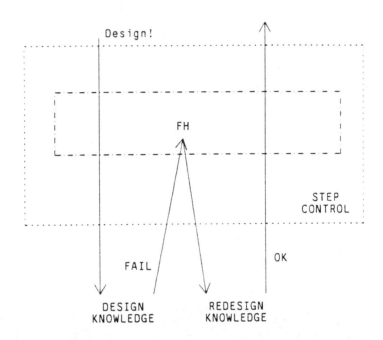

Figure 16: The Step's Redesign Action

If a step fails the appropriate FH is given the failure information. Usually this leads to a request for redesign. If there is no available information about redesign or if redesigning is known not to be possible then the FH will return a failure message. The FH will collect the suggestions from the failing step and remove all those that do not apply to attributes of steps over which the task has control. The remainder are kept to pass up to higher agents should the task fail in its redesign attempt. If there are no relevant suggestions then redesign cannot be attempted.

The next thing to do is to note the steps already executed by the task and to start to use a redesign strategy to control the redesign process. At this point we will only be discussing the "Least Backup by Suggestion" strategy as we feel that at the task level this is the one that is natural to use, as it is suggestion-driven and takes into account the step order, which in turn reflects inter-step dependencies. Another reason it is appropriate is that a task manipulates only a few attributes and design appears to be local "juggling" of these values until success is achieved. Other strategies will be discussed briefly in Section 4.5.3.4.

4.5.3.1 Task Backup Strategy

The Least Backup by Suggestion (LBBS) strategy uses the suggestions that refer to the steps already executed, and attempts redesign by picking suggestions in order of least backup. For example, if (A B C D) are the steps already executed, where D is the most recently executed, and there are suggestions that refer to C and A, then redesign of C will be tried before A, and A will only be tried if C fails.

The strategy starts by associating suggestions with the steps already executed. While all of these suggestions being considered refer to attributes over which the task has control, there may be none that refer to steps already executed. If that is the case, or if there are no steps already executed (i.e., the failing step is the task's first) then the redesign cannot continue and will fail.

For each of the steps already executed that have been suggested the strategy continues by requesting a redesign. Suppose A was suggested by C failing in steps (A B C), then A will get asked to do a redesign (see Figure 17). If that is successful the the intermediate steps (i.e., B) will be tested to see if they are affected by the changed attribute. If they are, a re-design is requested. If those are successful the failing step is asked to do a re-design. If that succeeds the normal design process continues using the task's remaining steps.

If a redesign request fails, or if an intermediate re-design request fails then the task redesigner will take the next least-backup step suggested by the original step failure and follow the same process. If all fail then the redesigner fails. If after the backup and redesign the failing-step fails again the simplest strategy is to fail and let the task redesigner try the next suggestion[15].

There are some more complex variants to this least-backup strategy. They all involve pursuing failures during the redesign process and attempting to fix them before continuing. The problem with them is that they lead to arbitrary levels of complexity. Certainly in practice, a designer, aided by pencil and paper to keep track of these

---

[15]This is the approach used in the implementation.

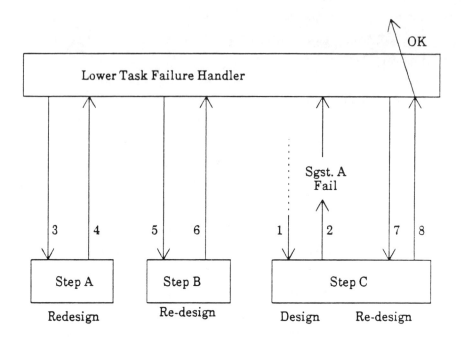

Figure 17: The Task Backup Strategy

levels and the intermediate status of the design, would be able to handle this complexity to a certain extent, but exactly to what degree is unclear.

4.5.3.2 Reduction of Suggestions

The task redesigner will ask steps below to redesign, passing them one or more suggestions. Those suggestions have been selected because they refer to the attribute over which that step has control. There can be more than one suggestion about an attribute as both the constraint and the step that failed can make them. In general, there could be any number of suggestions for some agent. Just prior to activating the redesigner the suggestions are reduced to a single representative suggestion that will "include" the others.

If any two suggestions "clash" (e.g., one says "increase" and the other "decrease") then the suggestions are incompatible and redesign cannot be done. Rules are applied until only one suggestions remains. For example, (INCREASE xyz BY 0.0015) and (DECREASE xyz BY 0.056) will result in failure; (INCREASE xyz BY 0.0055) and (INCREASE xyz BY 0.01) will result in (INCREASE xyz BY

0.01); (INCREASE xyz) and (INCREASE xyz BY 0.3) will result in (INCREASE xyz BY 0.3).

4.5.3.3 Pruning Suggestions

Note that both design knowledge and redesign knowledge are the result of experience. For some failures we have knowledge about what not to try during a recovery attempt. Some suggestions can be dismissed out-of-hand from experience as never working. This type of pruning results from a set of "ignorable suggestions" being associated with a task's failure handling and redesign. Any suggestions that arrive at the task after failure below can be ignored if they are in the set.

A more general form of pruning would involve more knowledge. For example, some knowledge can be used to check the size of the increment. This knowledge could equally well be in the redesigner itself so it is very difficult to justify any pruning that simply inspects individual suggestions using knowledge. If the knowledge uses a record of the prior instances of attempts to follow this suggestion, pruning could dynamically adjust to the performance of the redesigner. This requires more research.

4.5.3.4 Other Strategies

It seems quite clear that the underlying form of backtracking at the task level is **Least Backup by Suggestion**. However, we make no claims about it being the only strategy, or that it is always used in its pure form. Other strategies could be used both here and at the plan level where things are less clear. This needs more research to study the strategies actually used during design.

The most obvious backup strategy is **Chronological**. This would attempt redesign of each step in strictly reverse chronological order, i.e., most recent first. As this totally ignores any information gleaned from failures this is only a strategy of last resort.

Another possible strategy is **Most Suggestions First**. This strategy follows the reasoning that the more suggestions there are for altering an attribute the better its chances are of success and the sooner it should be tried. At the task level there are only likely to be at most two suggestions about an attribute -- one from the failing constraint and one from the step that contains it -- so at this level it is not quite so attractive as it first seems.

As the strategy too can be based on experience it could be a **Probability Ordered** strategy. It is reasonable to consider each suggestion that will occur being associated with some probability of success. The suggestions the task receives would be ranked in the task according to their probability of success, and tried in order.

It is difficult to believe that a designer has complete knowledge of this kind, but it is clear that he or she may have some ability to estimate probabilities. More likely is that the designer has some incomplete partial ordering of suggestions and will be able to produce a preferred order in some but not all situations.

4.5.3.5 Suggestions from the Task

On exit from a failing task an attempt is made to suggest how the problem can be fixed. These suggestions will be included in the failure message from the task along with all the suggestions that the task wasn't able to deal with because they refer to steps over which it has no control.

At the task level it is possible that rather than refer to attributes the suggestions should refer to collections of attributes. Thus a task's suggestion might be to (INCREASE PistonSealSeat) or (INCREASE Piston). Any agent receiving a suggestion such as this would have to know if the suggestion was relevant for it. Thus the task responsible for designing the Piston's Seal Seat would be able to recognize it and expand the suggestion into suggestions about certain attributes.

Notice that not all of the attributes of a component need to be increased in order to increase the whole. For example, to increase the size of a pencil one need not increase the diameter of the lead. Consequently suggestion expansion is dependent on knowledge that is local to the agent which recognizes the suggestion.

4.5.3.6 Revisions and Alterations

On entry to a task the state of the design is noted so that if it should fail all alterations made can be discarded. If the task succeeds then it requests a revision to the design, and all the alterations requested by the task's steps are noted in the revision. During failure handling the alterations that the task is requesting must be carefully monitored during any redesigns and re-designs.

4.5.4 Failure in a Specialist

A specialist can fail in several ways -- roughly, from within, and from below. Failures "within" are due to constraints at the level of the specialist. Failure from within may occur on entry, during plan selection, during plan execution, or on exiting from the specialist. Failure of the actions in the plan constitute failures from "below".

4.5.4.1 Specialist Failure on Entry

Specialist failure on entry is a sign that conditions are in some way inappropriate for this specialist to act. For example, a specialist may only be able to design air-cylinders smaller than a bread-box. On entry to a specialist, prior to plan selection, one would expect an examination of the gross conditions of applicability, and if those are not met a failure message would be sent to the calling agent.

4.5.4.2 Specialist Failure during Plan Selection

As has already been discussed, failure during plan selection can occur because:
- there are no plans appropriate for this situation;
- all appropriate plans have been used during this call of the specialist;
- there are no plans for this mode of use (e.g., Rough-design) of the specialist.

The second kind of plan selection failure is more complex than the others. It implies that all appropriate plans failed and that any local attempts to remedy the problems causing plan failure also failed. Clearly the calling agent should be informed of this special situation, as it represents the complete failure of the specialist.

4.5.4.3 Plan Failure

Failures "below" are failures of any action in the selected plan. Currently the possible actions in a plan are to call a specialist in design mode, call a specialist in rough-design mode, to execute a task and to test a constraint. We will discuss each of these in turn, and then discuss how a whole plan can fail.

Calling a specialist in design or rough-design mode are both very similar. Calling a specialist will be explicit in a plan, whereas calling for redesign will not be explicit in the plan proper, but is part of a specialist's information about how to react to failures. A design specialist can return the following messages:
- successful completion of design.
- failure due to no applicable plans.

- failure on entry.
- failure on exit.
- failure due to all plans failing.

In each case the message and action taken is recorded with the history of the action taken for this plan, which in turn is part of the history for this invocation. The message indicates that the called specialist has terminated.

In the case of a failing rough design call to a specialist the situation is similar. No redesign will be necessary. As this is attempting to prepare for the design proper, failure will assure that the design phase would fail. Failure will be reported to the calling specialist.

If the plan fails due to a failing constraint or to a failing task they are treated similarly. Both produce suggestions. The main difference is that, if recovery is possible the redesign strategy will send the failing task a re-design message, whereas the constraint will merely be retested.

A plan is an ordered sequence of actions, where the different kinds can be mixed in any way. In order to be successful, all of the actions must succeed. Note that this doesn't imply that nothing failed at a lower level, but merely that eventually every action in the plan returned success.

4.5.4.4 Specialist Backup Strategy

If an item of a plan fails then the specialist must consider the possibility of attempting to recover from the failure or giving up and causing the whole plan to fail. At the specialist level the recovery strategy is potentially more complex as there is more information to consider and more at stake. As each specialist can control different amounts of the design it should be clear that the detailed failure recovery activities for different specialists are likely to be different. Failure recovery within a plan should be carefully distinguished from plan selection after plan failure.

There are many pieces of information that might be included in the decision to try to recover, and the choice of strategy. Just as with task redesign it is possible that no single simple strategy is used by a human designer, but some combination according to circumstances. As the plan may contain specialists a lot of design may already have taken place prior to the failure. The more that has already been done, the greater the desire to minimize the amount to be changed. Thus, if a task can be asked to redesign in order to recover from the failure, this is much to be preferred over asking a specialist.

For some specialists a reasonable strategy is not to attempt to recover and merely to cause the plan to fail, leading to selection of another plan. The position of the specialist in the hierarchy, the degree of completion of the current plan, and the amount of backtracking suggested will all affect the decision whether to try redesign.

Another way of deciding whether immediate failure is appropriate is to use some measure of the "complexity" of the part of the plan already executed. One method of estimating complexity is based on an estimate of how many design agents it will use. If the complexity measure is too high then failure might be appropriate as too much needs to be altered. Conversely it could be argued that as a major investment in effort has been made it makes sense to attempt to recover if it is possible with a small amount of change to the existing design.

A more reasonable way of deciding whether to fail immediately is by using the notion of "key items". It is clear that when a designer is using a plan it is possible for him or her to realize that one or more parts of the plan (e.g., a task) is "key" in some way. That is, if it fails then there is no point proceeding with the plan. The agent that checks the requirements is one such key agent, as are all rough-design requests of specialists. This information is associated with each plan.

The Least Backup by Suggestion strategy is still very suitable in the specialist context. Those plan items which control attributes that have suggestions referring to them are tried in most recently used order. This is an intelligent backup strategy, but it can be improved.

Just as in the earlier discussion, strict chronological backup is not applicable here, but is mentioned for completeness. Using arguments similar to those used earlier, other reasonable strategies are possible, although we should be careful to avoid any theory that assumes that designers have complete knowledge of the situation.

4.5.4.5 Affected by Redesign?

After a plan item has been selected for redesign and it succeeds, the intermediate items (i.e., those between the redesigned item and the failing item), need to be asked if they are affected by the change. Consequently, specialists and tasks need to be able to answer that query. What is required is knowledge of whether the changed attribute (or attributes) is used as "input" to any of the steps over which they have control. This is not something a specialist or task can reasonably be expected to know. Consequently, in general the query will have to be passed down to the steps below until either one claims that it is affected or that all have been tried. In some cases, more reasonably at the task level, it can be expected that continued use would provide prestored answers to some but not necessarily all such queries.

4.5.4.6 Suggestion Pruning

At the specialist level, just as at the task level, there may be suggestions that should be ignored. This can be done by keeping a list of those suggestions to ignore. Through repeated application of failure recovery strategies it is probable that a finer evaluation of suggestions than merely include or ignore would be formed. At least, the suggestions would be formed into additional promising/not-promising categories. It is possible that the suggestions could be partitioned in this way prior to the application of some strategy, with the promising suggestions being tried first, and then the others. This is a diluted, but more realistic, form of a probability-based strategy.

4.5.4.7 Redesign Requested From a Task

If a suggestion from a failing plan item refers to some attribute over which a task in the plan has control then the redesign strategy could ask that task to redesign. For example, in the plan <T1 ; T2 ; T3> with T3 failing, T1 may be asked to do a redesign (see Figure 18).

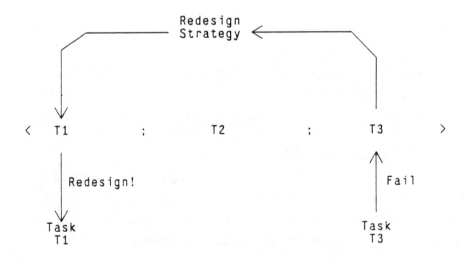

Figure 18: Redesign Request of a Task

There may be several steps below T1, and consequently, several attributes over which it has control. Consequently there may be more than one appropriate suggestion for T1. All of these suggestions should be sent to the task, and its redesigner will handle the suggestions appropriately using its local knowledge, and its strategies.

Note that these suggestions represent alternative ways to repair some failure in the failing plan item, and only one of them needs to be successful.

Notice that the task redesigner needs to return its success message, the suggestion that worked, the ones that failed, and those not yet tried. If the task's redesign leads to the failure of some intermediate plan item that is dependent on it (such as T2 above) then the task can be asked to try those suggestions which it has not yet tried.

4.5.4.8 Redesign Requested From a Specialist

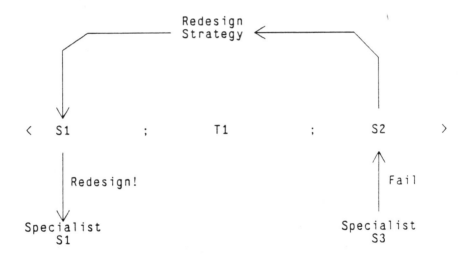

Figure 19: Redesign Request of a Specialist

Suppose the plan <S1 ; T1 ; S2> fails at S2 and the specialist S1 is suggested (see Figure 19). As above, there may be more than one suggestion that applies to attributes over which S1 has control, and, consequently, S1 will get passed all of those suggestions. The redesigner for the specialist S1 will need to select an appropriate redesign strategy.

As it is possible that different plans lead to different values for an attribute, the specialist must have a local memory of which plan was used successfully last time. Once the plan that was used last time has been found the redesigner can employ a redesign strategy, until one of the suggestions succeeds. The successful, failing, and unused suggestions should be returned with the success message, so that, if necessary, all suggestions can be attempted.

4.5.4.9 Re-design of Failing Plan Item

If after using some redesign strategy a failure goal has succeeded, the plan item that failed can now be re-designed. If the item is a task then not much effort will be required, and, on average, only a couple of step's activity will be duplicated. However, if the item is a specialist, much of the specialist's plan may have been successful before a failure occurred, even though it couldn't recover from it. In some cases it may be possible to keep some of the prior design.

4.5.4.10 Suggestion-to-Plan-Item Matching

When a failing plan item makes suggestions there must be some way of deciding which of the previously executed plan items are being implicated and which suggestions do not apply here and should be saved to be passed upward on failure of the recovery attempt. One possibility is to assume that the specialist has this knowledge. That is, for every attribute controlled by some item in one of its plans, the specialist can say to which plan item the suggestions about that attribute should be directed. For specialists deep in the hierarchy (i.e., near or at the tip nodes) this kind of knowledge is plausible. However, in general each specialist may have to make this mapping for hundreds of different attributes.

It is clear then that a better solution is to have every plan item decide which suggestions apply to it. Tasks will immediately recognize which suggestions apply to it, whereas specialists will have to query the items of the last successful plan. Of course, with continual use, some of this querying process may become short-circuited and the specialist would immediately recognize some attributes as belonging to certain plan items.

4.5.4.11 Plan Selection after Failure

Plan selection has already been discussed. After a plan fails another needs to be selected. If there are no more to select from, plan selection will fail. Available failure information includes the list of plans already tried and the reasons for their failure. The selector could incorporate rules such as "If plan A fails don't bother trying plan B", or "If both plan A and plan B failed plan C should be tried".

4.5.4.12 Suggestions from a Specialist

On exit, a failing specialist may make suggestions. It is unlikely that there will be many suggestions that a specialist can make about a single attribute. It is more likely that the suggestions will be of the group-of-attributes kind described above. Some specialists will not make suggestions.

4.5.4.13 Alterations, Revisions and Updates

On exit from a failing plan the state of the design is returned to its state on entry. This is done by undoing all of the updates made to the drawing from within this plan, thus reverting to an old drawing. On exit from a successful plan an update is made. As each failing step takes care of its alterations, and each failing task takes care of its revisions, this is all that needs to be done.

4.6 Failure-handling in Other Research

4.6.1 Styles of Failure Handling

For those used to programming, the natural way to think of dealing with failure in some part of a process is to backtrack chronologically along the trace of control flow until some place is reached where there is an alternative course of action (i.e., a choice point) and then make some new choice. This is the PLANNER [Hewitt 72] [Sussman 71] method of dealing with failure, although it was augmented by a limited failure message capability that would allow some information about the failure to be returned. With the development of CONNIVER [Sussman 72] the user was able to choose whether to do backtracking or not. In addition it was possible to inspect, destroy or reactivate suspended routines that had been kept with their environments, allowing arbitrarily complex control schemes. The data-base 'context' mechanism made data-base manipulations and recovery at failure time much easier, as each choice could be associated with a separate layer of the data-base.

Dependency-directed backtracking provides a way of taking into account more of the nature of the failure than chronological backtracking does. Dependency records are kept to record the way that the use of givens, assumptions and deductions depend on prior information. One use of this mechanism is to make "out" any facts in the data-base that are no longer to be believed as a result of a failure (contradiction). Processing concentrates on those items in the data-base that are "in" (i.e., believed). Additionally one can record the contradiction situation separately from the dependency information. In a rule-based system this leads to a new problem situation that will trigger new rules.

Another way of using dependency information is to allow the information about which pieces of knowledge contribute to the failure (assertions or actions) to be used in the decision of how far to backtrack. Nodes in the dependency structure are associated with their sources, and the use of these sources is noted in a chronological trace of the control flow. Consequently, at failure time, backtracking can be done to the last contributing choice point, as opposed to just the last choice point. A clear and concise introduction to and bibliography for techniques such as the above, referred to as "belief revision", can be found in Doyle [Doyle 80].

4.6.2 EL/ARS

The EL/ARS system for electronic circuit analysis [Sussman 75] [Stallman 77] [Sussman 77] is an example of a system that uses dependency records to keep track of failure situations. As it is "demon"-driven it appears to do no control backtracking, but avoids failure situations in further processing by using all of the recorded "NOGOOD" situations to affect the choice (i.e., triggering) of actions. As not all NOGOOD situations are relevant for any one problem-solving situation this method is missing some essential structure and can be inefficient. The use of 'pure' dependency structures in a global way (i.e., across the whole problem) is also an example of unconstrained use of a technique in a way that doesn't reflect the structure of the problem-solving activity.

4.6.3 TROPIC

Latombe's design system TROPIC [Latombe 76] [Latombe 79] uses dependency directed backtracking to the most recently contributing choice point. Failure information is attached to the chronological trace of control flow at a point singled out at backup time as being responsible for the failure. This is clearly better than pure chronological backtracking, and the controlled use of failure recording is to be preferred over the EL/ARS method. However, there is no control over the range of the backtracks made, as the system maintains a global dependency structure.

4.6.4 DESI/NASL

McDermott's approach to failure handling is different. In his DESI/NASL system [McDermott 77] [McDermott 78] failure handling is treated as just another task for the system to handle. The system is prevented from backing up over a previously made decision about a value. McDermott argues that, in general, backing up to a choice point using a universal mechanism is not appropriate, and that consideration must be given to all the actions and choices made since that point and prior to the failure. The failing primitive task produces a description of the failure, and a failure task is set up to attempt to deal with the situation. As well as that description, the task/subtask structure, the control trace and the recorded data dependencies are available for use. New subtasks may be added, old subtasks restarted, old subtasks re-expressed, or heuristics abandoned.

4.6.5 BUILD

In the BUILD planning system Fahlman [Fahlman 73] adopts an approach similar to McDermott's. Fahlman wrote the system in CONNIVER using control structures that allowed "the BUILD-PLACE-MOVE sequence to proceed in a headlong manner, with very little pre-checking of conditions", and reported that "trouble is met in a variety of ways when it arises" [p.117]. The suggestion here is that failure is normal, and that for unfamiliar situations the plans used are often only "almost right". This is close to the notion of debugging almost right plans that subsequently appeared in the work of Sussman [Sussman 75] and Goldstein [Goldstein 74].

In the BUILD system, every function that makes a major choice includes the declaration of a "Gripe Handler". If the subgoal selected fails in some way the most local gripe handler is called with a failure message reporting the problem. The gripe handler has access to the full environment of the failure situation, and can also examine the bindings of the choice function. Should the decision be made that failure is due to some decision at a higher level it may complain to the next higher gripe handler. If the problem is local it can suspend that subgoal, carry on in some way, or do chronological backtracking.subgoal.

The problems with this method of failure handling are the same as for the DESI/NASL system. That is, there are no constraints on the backtracking done.

4.6.6 MEND

The MEND system of Srinivas [Srinivas 78] attempts to recover from execution errors in robot plans by determining the cause of the failure and devising a recovery plan. Analysis of failure is thought of as "limiting the set of all possible explanations to the specific one that applies in a particular situation" (p. 276). On failure, a tree is built back from the failure point to include the previous actions and all the possible contributing reasons for failure. Some of these branches can be eliminated.

The possible reasons for failure, and the information about distinctive features are associated with each type of action, as is information about how to correct errors local to that action. Some errors, for example, those due to incorrect information or failing preconditions, will be fixed by undoing certain actions, reachieving preconditions or getting the necessary information, and redoing the undone steps.

Srinivas has effectively preclassified all failure situations and allowed the system to build just that part of the failure explanation tree local to the failure. As there is a fixed number of actions each with its own goal and explicit or implied pre- and post-conditions this kind of analysis can be done. In our work the situations and goal for each step is unique. Even though in some sense each failing agent has explanations attached it is not possible to say exactly which one is the actual reason or to specify a single recovery method.

4.6.7 Other research

Failure Handling is also addressed in some planning and robotics systems. The planning and plan execution systems of Nilsson [Nilsson 73] and Hayes [Hayes 75] both include the notion of 'failure handling'. Nilsson's system is a simple hierarchical planner that mixes planning steps with execution in a way that handles failures by chronological backtracking. Hayes describes a similar system that maintains a goal/subgoal structure along with the reasons for subgoal production. If plan execution leads to failure then the 'reasons' are used to assist in replanning.

The HACKER planning system [Sussman 73] used information about failures in plans to improve them. Knowledge about failures was encoded in Critics, that watch for occurrences of known types of bugs in plans as they are being developed, and in a plan simulator that watches for failures during simulated plan execution.

In Robotics the work of Gini and Gini [Gini 83] specifically addresses failure handling while executing robot programs. Their approach has many similarities to ours. They use Sensor Interpretation Rules in order to discover what error occurred (similar to Failure Handlers) and Recovery Rules to control the return to a normal state. In their use it appears that the recovery rules specify exactly what to do to recover from failure, and that all the failure recovery methods are explicitly stated.

Schank [Schank 77] describes the use of Scripts to understand stories about sequences of situations with "interferences" in them -- that is, "states or actions which prevent the normal continuation of a script" (p.52). He describes "obstacles" which remove some enabling condition for some action, and "errors" where an action produces an unexpected result. These are akin to constraint checking on entry to an agent, and checking after some decision about a value. Corrective actions, often "loops" which are repetitions of the failing action with some prior change, can be made or the script can be abandoned. Loops correspond to redesign attempts.

4.7 Summary

This chapter presented the types of knowledge necessary to handle the failures that occur during design problem-solving. Failure Handlers and Redesigners were introduced. Failure recovery action and the knowledge involved was presented for each agent. The role of Suggestions and Redesign Strategies were discussed. The handling of Plan failures was also presented. The chapter concludes by surveying some of the other methods of failure handling described in the literature.

In the next chapter we present DSPL, a language that allows expression of knowledge about how to do a routine design. In DSPL one writes in terms of the types of agents, such as Specialists, already presented in Chapter 3.

5 DSPL: A Language for Design Expert Systems

5.1 Introduction

A trend during recent years has been to produce Expert System building tools. The EMYCIN system, for example [van Melle 79], is MYCIN [Shortliffe 76] "stripped of its domain knowledge". Many other tools are available (see [Hayes-Roth 83] Chapter 9).

These ES building tools have only been partially successful in providing facilities explicitly tailored to one type of problem solving. Most of them, due to reliance on relatively unstructured collections of rules, do not control the type or manner of problem-solving activity very strictly. By using the theory of generic tasks (see Chapter 1) it is possible to produce ES building tools which constrain and guide the user toward building an ES from a combination of types of problem-solving. This approach has been followed in the CSRL language for Classification and Hypothesis Matching [Bylander 83], the IDABLE language for Knowledge-Directed Information Passing [Mittal 84] [Sticklen 83], and in the DSPL language presented here.

5.1.1 The Design Specialists and Plans Language

DSPL provides a way of writing declarations of Specialists, Plans, Tasks, Steps, Constraints, Failure Handlers, Redesigners, Sponsors and Selectors, allowing the user to specify the knowledge contained in them. In the following section we will address each of these declarations in turn.

To build a Design Expert System (ES), the user declares in DSPL all the agents required, and then allows the underlying DSPL system to link them together after some checking. Once formed the ES can be invoked by requesting a design from the top-most specialist. The design then proceeds according to the specialist's plans. After a successful completion the design data-base contains the completed design. If failure occurs reasons are given. The DSPL system provides the underlying problem-solving control.

As the DSPL system is implemented in Lisp, the language has a very Lisp-like form for implementation convenience, but there is no reason in principle why it should be tied to such a form.

5.1.2 DSPL Conventions

The conventions used in DSPL are as follows:

- all the words provided by the language are in upper-case (e.g., NAME, PLAN);
- where possible things are presented in an attribute/value form, where the attribute is some system defined name;
- every declaration in DSPL of an entity associates a name with the entity, and references to other entities (e.g., steps in a task) are by using their names;
- calculations are expressed by using specially defined functions that allow tolerances to be manipulated;
- all variables used in an agent, for holding values from the design database or from calculations, are considered to be local to that agent and exist for the lifespan of the agent;
- assignment to a variable is expressed by writing a variable name and then the expression whose value is to be assigned.

In the following sections each major declaration available for a user will be presented using a typical example taken from the AIR-CYL system. They are intended to present major features of the language only and not be an exhaustive presentation. The language is still being developed and refined. We will not discuss every part of every example. In many cases default declarations are used by DSPL if information is missing. A detailed description of the syntax is given in Appendix E.

5.2 Specialist Example

```
(SPECIALIST
  (NAME Head)
  (USED-BY AirCylinder)
  (USES None)
  (DESIGN-PLANS HeadDP1)
  (DESIGN-PLAN-SELECTOR  HeadDPSelector)
  (ROUGH-DESIGN-PLANS HeadRDP1)
  (ROUGH-DESIGN-PLAN-SELECTOR  HeadRDPSelector)
  (INITIAL-CONSTRAINTS None)
  (FINAL-CONSTRAINTS None)
)
```

Figure 20: Specialist "Head"

If no SELECTOR is specified a default selector will be used which selects PLANS in declaration order. Throughout the examples a fairly strict naming convention has been adhered to. For example, HeadRDP1, which stands for "first rough design plan for the head". The system does nothing to check or support this. However, such a scheme is essential for expressing large amounts of knowledge in a language such as DSPL.

5.3 Plan Example

```
(PLAN
  (NAME HeadDP1)
  (TYPE Design)
  (USED-BY Head)
  (SPONSOR HeadDP1Sponsor)
  (QUALITIES Cheap)
  (INITIAL-CONSTRAINTS None)
  (FINAL-CONSTRAINTS None)
  (BODY HeadTubeSeat
      MountingHoles
      Bearings
      SealAndWiper
      AirCavity
      AirInlet
      TieRodHoles
      (REPORT-ON Head)
) )
```

Figure 21: Plan "HeadDP1"

The TYPE of a plan can be "Design" or "RoughDesign". The QUALITIES are given in order of importance. At present, however, plan selection does not retrieve these directly from the plan but considers them to be built into the sponsor's knowledge. If no SPONSOR is specified then a default sponsor is used. This always considers the plan to be perfect, unless it has already been tried. The BODY contains the details of the plan, and consists of an ordered list of plan items. In this example the plan consists entirely of tasks, with the exception of the last item which is a function provided by DSPL to print out the attributes and values of some part of the design.

5.4 Task Example

```
(TASK
  (NAME AirCavity)
  (USED-BY HeadDP1)
  (BODY
    AirCavityDepth
    AirCavityID
    AirCavityOD
    CheckAirCavity
))
```

Figure 22: Task "AirCavity"

A task can be USED-BY more than one plan. The REDESIGNER, as not specified, will default to a system version. Two system FAILURE-HANDLERS are also provided by the system. The INITIAL-CONSTRAINTS and FINAL-CONSTRAINTS will both default to "None". The BODY of the task consists of a sequence of steps, referred to by name.

5.5 Step Example

The sample step is USED-BY the AirCavity task. The ATTRIBUTE-NAME is the attribute for which this step is to design a value. If a failure occurs the REDESIGNER will attempt to recover from it. The declaration REDESIGN NOT-POSSIBLE is also allowed. If the step itself fails the FAILURE-SUGGESTIONS get passed up with the failure message. Each item in the suggestion list is evaluated at failure time. This allows conditional suggestions such as (IF (> x y) THEN (SUGGEST ...)). If the suggestion includes an expression, as in (DECREASE xyz BY (+ pqr 0.56)), then the SUGGEST function will arrange for the value to be computed. The actions DECREASE, INCREASE or CHANGE refer to attributes. In the current system all attribute names must be unique.

The BODY of the step is divided into KNOWN and DECISION sections. Singular or plural keywords may be used as required. The KNOWN section obtains the values from the design data-base by doing KB-FETCH. The KB-FETCH uses the component and attribute names. The single quote (i.e., ') is used to indicate that the name given is to be used directly without evaluation, as opposed to the use of a variable (e.g., HeadMaterial) that should be evaluated prior to use (e.g., giving, for example, Aluminum).

```
(STEP
  (NAME AirCavityID)
  (USED-BY AirCavity)
  (ATTRIBUTE-NAME  HeadAirCavityID)
  (REDESIGNER AirCavityIDRedesigner)
  (FAILURE-SUGGESTIONS
    (SUGGEST (DECREASE RodDiameter))
    (SUGGEST (DECREASE HeadBearingThickness))
    (SUGGEST (CHANGE HeadMaterial
             TO DECREASE MinThickness))
  )
  (COMMENT "Find air cavity internal diameter")
  (BODY
    (KNOWN
      BearingThickness (KB-FETCH
                'Head 'HeadBearingThickness)
      RodDiameter (KB-FETCH 'Rod 'RodDiameter)
      HeadMaterial (KB-FETCH 'Head 'HeadMaterial)
      MinThickness (KB-FETCH
              HeadMaterial 'MinThickness)
    )
    (DECISIONS
      MaxRodRadius       (VALUE+ (HALF RodDiameter))
      MaxBearingThickness (VALUE+ BearingThickness)
      AirCavityRadius    (+ MinThickness
                 (+ MaxRodRadius
                    MaxBearingThickness))
      AirCavityID  (DOUBLE AirCavityRadius)
      REPLY       (TEST-CONSTRAINT ACID)
      REPLY       (KB-STORE
              'Head 'HeadAirCavityID AirCavityID)
    )
) )
```

Figure 23: Step "AirCavityID"

The DECISIONS section consists of variable-action pairs, where the action is evaluated and its value assigned to the variable. That variable may then be used in subsequent actions in the step. The variables assigned values in the KNOWNS section may also be used. Arithmetic expressions use prefix operators. The function VALUE+ returns the value plus the positive tolerance of the value, and consequently provides the largest magnitude for that length. There are many other functions available.

There are two distinguished variable names. One is REPLY, the other COMMENT. A comment acts as a dummy assignment and expects a string of text as the "action". This is just a way of inserting a comment into the body of a step. A REPLY variable is used when no value is produced by the action but a message showing success or failure is produced instead. The TEST-CONSTRAINT and the KB-STORE are two examples. Failure will stop the execution of the body and will cause the DECISION section to fail. The value calculated by the step is put into the design data-base with a KB-STORE.

5.6 Constraint Example

```
(CONSTRAINT
  (NAME ACID)
  (USED-BY AirCavityID)
  (FAILURE-MESSAGE
   "Air cavity ID too large relative to tube ID")
  (FAILURE-SUGGESTIONS
    (SUGGEST (INCREASE TubeID
          BY FAILURE-AMOUNT))
    (SUGGEST (DECREASE HeadAirCavityID
          BY FAILURE-AMOUNT))
  )
  (BODY
    (KNOWN
     TubeID  (KB-FETCH 'Tube 'TubeID)
    )
    (TEST
     ( AirCavityID
       <
       (TubeID - (DOUBLE MinThickness))
    ) )
) )
```

Figure 24: Constraint "ACID"

The FAILURE-MESSAGE, in the form of a string, is a unique message that will identify this constraint if it fails. Clearly it is preferable to have the message have some meaning for the human reader, although it need not. It does, however, need to be recognized by a Failure Handler. The FAILURE-SUGGESTIONS are used if the constraint fails. Notice that the suggestions include BY FAILURE-AMOUNT. The variable FAILURE-AMOUNT will be set by the system at the time of constraint failure to the amount by which the TEST failed. In fact, any variable or arithmetic expression can be used after the separator BY.

The KNOWNS section of the BODY gathers all the information needed to make the test, and may include some calculations. The TEST consists of a single relation in infix notation, possibly including arithmetic expressions. The variables in the test may include not only those from the KNOWNS section but also any that are active in the agent using the constraint (e.g., MinThickness and AirCavityID).

5.7 Redesigner Example

The ADJUSTMENT is the smallest machinable change for that attribute and will be provided by the system if the user doesn't specify it. The VALUE acts as a local variable that holds the current value of the attribute being redesigned. The value will be obtained from the design data-base, or locally from within the failing agent, as appropriate.

Instead of a single BODY, a redesigner has INCREASE, DECREASE and CHANGE action sections, any of which may be omitted or declared as being not possible (e.g., CHANGE-NOT-POSSIBLE). Each action section has its own KNOWNS and DECISION sub-sections. Each action section will respond to a matching type of suggestion (e.g., increase) from above.

In each decision sub-section the variables VALUE and INCREASE, DECREASE or CHANGE are used to produce the suggested alteration {e.g., (- VALUE DECREASE)}. In the DECISION sub-section of the DECREASE action section notice the use of an IF...THEN...ELSE construct. Note that it will be expected to return success or failure as its result has been assigned to a REPLY variable. The DECREASE-NOT-POSSIBLE variable is preset to a failure message. The two main forms of redesign action sections are shown by the INCREASE and the DECREASE of this example.

```
(REDESIGNER
  (NAME AirCavityIDRedesigner)
  (USED-BY AirCavityID)
  (USED-BY-TYPE Step)
  (ADJUSTMENT 0.001)
  (VALUE (KB-FETCH 'Head 'HeadAirCavityID ))
  (INCREASE
   (KNOWNS
    TubeID     (KB-FETCH 'Tube  'TubeID)
    Material   (KB-FETCH 'Head  'HeadMaterial)
    MinThickness (KB-FETCH Material 'MinThickness)
    )
   (DECISION
    AirCavityID (+ VALUE INCREASE)
    REPLY      (TEST-CONSTRAINTS ACID)
    REPLY      (KB-STORE
         'Head 'HeadAirCavityID AirCavityID)
  ) )
  (DECREASE
   (KNOWNS
    TubeID     (KB-FETCH 'Tube  'TubeID)
    Material   (KB-FETCH 'Head  'HeadMaterial)
    MinThickness (KB-FETCH  Material 'MinThickness)
    BearingThickness
         (KB-FETCH 'Head 'HeadBearingThickness)
    RodDiameter (KB-FETCH 'Rod 'RodDiameter)
    )
   (DECISION
    AirCavityID (- VALUE DECREASE)
    REPLY      (TEST-CONSTRAINTS ACID)
    MinIDRadius (+ MinThickness
           (+ (VALUE+ (HALF RodDiameter))
              (VALUE+ BearingThickness) ))
    REPLY (IF ( >= AirCavityID (DOUBLE MinIDRadius))
        THEN (KB-STORE 'Head
            'HeadAirCavityID AirCavityID)
        ELSE DECREASE-NOT-POSSIBLE
  ) )    )
  (CHANGE  CHANGE-NOT-POSSIBLE)
)
```

Figure 25: Redesigner "AirCavityIDRedesigner"

5.8 Failure Handler Example

```
(FAILURE-HANDLER
  (NAME SystemStepBodyFailureFH )
  (TYPE System)
  (USED-BY-TYPE Step)
  (BODY
    (TABLE
      (DEPENDING-ON  MESSAGE)
      (MATCH
        (IF ( "Programming problem" )
         THEN  (FAIL) )
        (IF ( "Pre-decision constraint failure" )
         THEN  (USE-FH SystemPreDecisionConstraintFH
             WITH CONTRIBUTING-MSG) )
        (IF ( "Failure while getting known values" )
         THEN  (USE-FH SystemGettingKnownsFH
             WITH CONTRIBUTING-MSG) )
        (IF ( "Decision failure" )
         THEN  (USE-FH SystemDecisionFH
             WITH CONTRIBUTING-MSG) )
      )
      (OTHERWISE
        (DO (COMPLAIN "Message not recognized" )
          (FAIL)
      ) )
) ) )
```

Figure 26: Failure Handler "SystemStepBodyFailureFH"

A Failure Handler (FH) can be of type "System" or one of a set of user FH types. For example, a major type in the implementation is the USER-DECISION-CONSTRAINT-FH. This is responsible for recognizing and reacting to the failure of one specific constraint in a step's decision section. All user FHs will be defaulted by the system if they are not supplied.

The FH in this example will be responsible for recognizing the failures that emanate from the BODY of a step, and consequently the USED-BY-TYPE is "Step". Note that this FH declaration is one that is essential to the building of the system, but, once provided, will not need to be provided again -- i.e., it is domain independent. The user will not need to worry about writing System Failure Handlers at all, as they are already provided as part of the DSPL system.

The BODY of the FH consists of a single function. The TABLE provides a way of selecting actions based on the values of one or more variables. In the example the only variable is the system variable MESSAGE that is set to the message string identifying the current failure. The system will automatically assign the string to the variable. The MATCH section provides a number of alternatives for the message and specifies actions to be taken where a match is found. So, if the message was "Programming problem", a FAIL function would cause immediate failure of the FH. For other messages the action is to call some other FH and pass it the failure message that caused the currently examined failure (i.e., the CONTRIBUTING-MSG). Thus, eventually, the original cause of failure will be passed to an FH and recovery action can be specified. If no match occurs, the OTHERWISE part of the TABLE specifies what to DO, i.e., COMPLAIN and FAIL.

5.9 Sponsor Example

The NAME and which Selector it's USED-BY is declared, along with the name of the PLAN to which this Sponsor is attached. Note that unlike the other DSPL examples this and the Selector example below have not been taken from the AIR-CYL system's DSPL.

The BODY's Variable-Action pairs represent assignment. In a sponsor, the distinguished variable REPLY expects a Suitability and will terminate execution of the BODY if one is assigned, otherwise execution continues. The predicate ALREADY-TRIED? will check to see if the argument is a plan that has already been tried and has failed. The functions "something"-REQS will provide the values of these qualities from the requirements. The variable LAST-FAILING-ITEM provides the name of the failing item in the last failing plan. The function COMBINE will produce a suitability from the two calculated, according to a simple but slightly pessimistic method.

Note that the Sponsor is expected to provide a suitability. If it doesn't then a "use of plan language" failure will occur. Also, if a plan doesn't declare a sponsor then a Default Sponsor will be used whenever necessary and it will always report the plan as PERFECT, unless it has already been tried, in which case it will return RULE-OUT.

5.10 Selector Example

The various SELECTION-METHODS act as subroutines. In this case there is only one METHOD, with the NAME Method1, and an INPUT-VARIABLE PlanNames. The method will get used in the body of the selector. The function LEAST-COMPLEXITY will return the least complex plan(s) from those given to the method, and, if there is only one its name will be the result of the method. If not, one is selected from the remaining plans by virtue of the designer preferring that one, i.e., it is earlier in the

```
(SPONSOR
  (NAME   ExampleDPSponsor)
  (USED-BY ExampleSelector)
  (PLAN   ExampleDP3)
  (COMMENT "gives suitability")
  (BODY
   REPLY (IF (ALREADY-TRIED? PLAN) THEN RULE-OUT)
   REPLY (IF (ALREADY-TRIED? 'DP1) THEN RULE-OUT)
   Qualities
    (TABLE (DEPENDING-ON
             (RELIABILITY-REQS)
                 (MANUFACTURABILITY-REQS)
                     (COST-REQS) )
        (MATCH
         (IF (Reliable  Easy  Cheap)
          THEN PERFECT)
         (IF (Medium   Easy   ? )
          THEN SUITABLE) )
        (OTHERWISE RULE-OUT)
    )
   Situation
    (TABLE (DEPENDING-ON
          (KB-FETCH 'Part 'Width) )
        (MATCH
         (IF ( (> 6) ) THEN RULE-OUT)
         (IF ( (<= 6) ) THEN SUITABLE)
    )  )
   Agent   (EQUAL 'Task2 LAST-FAILING-ITEM)
   REPLY
    (TABLE (DEPENDING-ON
          Agent Qualities Situation)
        (MATCH
         (IF ( T  ?       ? )THEN RULE-OUT)
         (IF ( ? SUITABLE  ? )THEN SUITABLE))
        (OTHERWISE
          (COMBINE Situations Qualities) )
) ) )
```

Figure 27: Sponsor "ExampleDPSponsor"

designer specified plan order.

The BODY of the selector is in Variable-Action form. In a selector, the distinguished variable REPLY expects a plan name, and will cause termination of the execution of the BODY if one is assigned. The variables PERFECT-PLANS and SUITABLE-PLANS are automatically set to the names of the plans with those sponsor-given suitabilities.

```
(SELECTOR
  (NAME    ExampleDPSelector)
  (USED-BY Example)
  (TYPE    Design)
  (USES    ExampleDPSponsor SampleDPSponsor)
  (COMMENT "gives name of plan")
  (SELECTION-METHODS
   (METHOD (NAME Method1)
      (INPUT-VARIABLE PlanNames)
      (BODY
         LC   (LEAST-COMPLEXITY PlanNames)
         REPLY (IF  (ONLY-ONE? LC)
              THEN LC
              ELSE
                (DESIGNER-PREFERENCE LC)
)) )     )
  (BODY
   REPLY (IF (MEMBER 'Plan26 PERFECT-PLANS)
       THEN 'Plan26
       ELSE
        (IF PERFECT-PLANS
         THEN (USE-METHOD Method1
              ON PERFECT-PLANS)
         ELSE
          (IF SUITABLE-PLANS
           THEN (USE-METHOD Method1
                ON SUITABLE-PLANS)
           ELSE NO-PLANS-APPLICABLE
      ) ) )
) )
```

Figure 28: Selector "ExampleDPSelector"

The BODY starts by checking to see if Plan26 is one of the perfect plans, and if so it is selected immediately. Next Method1 is used on the list of perfect plans, but if there weren't any perfect plans Method1 is used to select from the suitable plans. If there weren't any of those, then it reports that no plans were applicable. The BODY is expected to return a plan name, and if it doesn't then a "use of plan language" failure will result.

5.11 Summary

This chapter has introduced the idea of languages in which to express problem-solving knowledge, and has presented DSPL for design problem-solving. Examples of Specialist, Plan, Task, Step, Constraint, Redesigner, Failure Handler, Sponsor, and Selector knowledge were given.

In the next chapter we present the AIR-CYL system as an example of the use of DSPL to build a routine design problem-solver.

6 AIR-CYL: An Air Cylinder Design System

AIR-CYL is a routine design system implemented using DSPL. The implementation of DSPL[16] for AIR-CYL was done on a DECsystem-20, written in Elisp (a version of Lisp from Rutgers University). An OSU version of FRL [Roberts 77] was used for the design data-base. The AIR-CYL system takes between 1 and 5 minutes to do a complete design with no redesign, depending on the load on the system.

We will present some comments about the Air-Cylinder domain, the AIR-CYL system itself, the major features of the DSPL system as it has been implemented, and a description of the Design Data-Base.

6.1 An Instance of Class 3 design

6.1.1 The Air-cylinder

Let us consider a real but not overly complex example. In our collaboration with AccuRay Corporation, an Air-cylinder (AC) was selected as a suitable object for our continuing studies of design problem-solving (See Figure 29). Our preliminary work on design problem-solving was reported in [Brown 83]. Since then we have been working on extending the theory and examining the issues and problems using the AC as our test case.

The AC has about 15 parts, most of which are manufactured by the company according to their own designs, as their requirements are such that the components cannot be purchased. The AC is redesigned and changed slightly due to applications with different requirements, and, consequently, the domain is Class 3 in type. In operation, compressed air forces a piston back into a tube against a spring. Movement is limited by a bumper. The spring returns the piston, and the attached "load", to its original position when the air pressure drops.

---

[16] A more recent version of DSPL has been implemented using LOOPS on a Xerox Lisp machine. Inquiries about obtaining a copy of DSPL should be addressed to Dr. B. Chandrasekaran at Ohio State University.

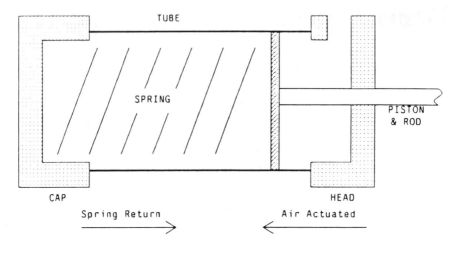

Figure 29: Air-Cylinder

6.1.2 The Conceptual Structure for the Air-cylinder

An Air-cylinder Designer was interviewed over a period of time, the protocols were analyzed and the "trace" of the design process was obtained. Figure 30 shows the progress of the design over time (from left to right) and the groupings of the decisions (from top to bottom).

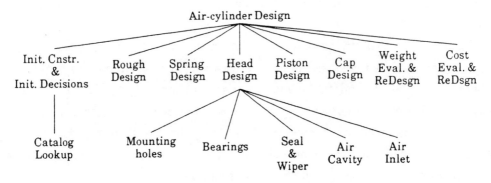

Figure 30: Design Trace

The trace was subsequently analyzed to establish the underlying conceptual structure. For example, the Head was clearly treated as a separate conceptual entity, as it occupied a substantial portion of the designer's time and effort. The Spring was actually designed by a different person as an essentially parallel activity, while the rest of the design was "lumped together" by the designer as the third major activity. The fact that the specialists can be fairly easily identified, and that the plans for each specialist are also identifiable and small in number, strongly confirms that this is a Class 3 activity. On examination we could see that this organization tends to localize dependencies, and allows for parallel design activity.

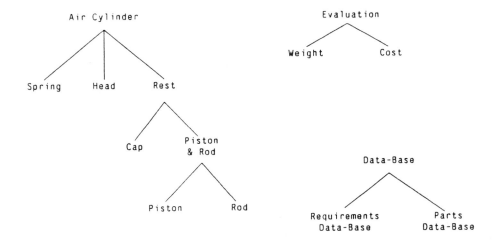

Figure 31: Partial AIR-CYL Structure

Figure 31 shows an outline of the conceptual structure believed to underlie the design problem-solving. Part of the specialist hierarchy is shown on the left of the diagram. The data-base needed during the design must keep track of how the design of the parts is progressing, and must keep the requirements available for continuous access by all the specialists. After the design is complete, and to a lesser extent during the design, an evaluation is done of weight, stress and cost. In a complete design system all these types of knowledge must be represented and must interact appropriately to produce the trace shown above.

6.2 The AIR-CYL System

6.2.1 Requirements Checking

There are 19 values provided as input to the system. They are Envelope Length, Envelope Height, Envelope Width, Maximum Temperature (degrees F), Operating Medium (e.g, Air), Maximum Operating Pressure (psig), Minimum Operating Pressure, Load in the Rod (lbs), Stroke, Rod Thread Type (e.g., UNF24), Rod Thread Length, Rod Diameter, Environment (e.g., Corrosive), Quality of design (e.g., Reliable), MTBF (hours), Air Inlet Diameter, Mounting Screw Size, Mounting Hole to Mounting Hole Distance, Maximum Face to Mounting Hole Distance. All lengths are in inches.

The system allows the requirements to be entered individually one at a time, but it also allows them to be read from a disc file. Currently they are expected to be expressed as an FRL frame, although this limitation can easily be improved.

The extent of the design knowledge is such that, even though all of the requirements are checked, not all of them are used in the design. This is due to the limited amount of debriefing of the designer and our ignorance about Air Cylinders. The designer was insistent that all of the requirements played some part in the design.

6.2.2 Rough Design

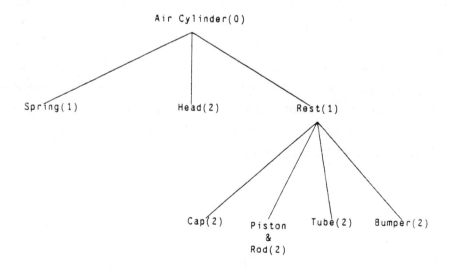

Figure 32: Rough Design Hierarchy

In Figure 32 the rough design hierarchy is shown. Notice that in this case the hierarchy is the same as the design hierarchy -- i.e., all of the specialists are involved. The numbers in parentheses show the number of tasks local to each specialist. These are specified in rough design plans, and are substantially fewer than for design plans. Note that most of the tasks have only a single step in their body, and that often what gets decided is a major dimension and/or the material. Rough design is entirely controlled by plans expressed in DSPL. Specialists called in RoughDesign mode use rough design plans.

6.2.3 Design

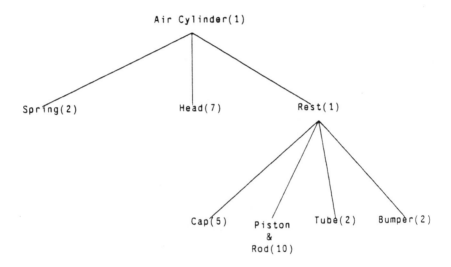

Figure 33: Design Hierarchy

In Figure 33 we show the design hierarchy used, along with the number of tasks local to each specialist. Each task will have several steps. The design process is entirely determined by the design plans that have been defined in DSPL for each specialist.

The major omission from the design knowledge is the complete design of the spring. The designer interviewed did not design the spring, and did not know about spring design. The system has preset in it the essential dimensions of the spring so that the complete design can be done. This has the added advantage that failure handling can be tested by forcing failures merely by shrinking any or all of the envelope dimensions. As the spring size is fixed, shrinking the Cap and Head will eventually lead to failures (see traces in Appendices B and C).

Many of the steps have been completed using common sense and some general mechanical knowledge, as the designer interviews didn't always proceed in all areas to the required level of detail. Given the same requirements that the designer was given the AIR-CYL system provides a very similar design to the one that the designer produced, except that it is a little more conservative than the designer in many places, as in many steps we enforced the suggested minimum thicknesses of materials. As differences in measurements affect other measurements it is difficult to compare the designs directly. For example, for the Head, the average "error" (i.e., the difference between the designers value and the system's) is about 5% if one includes two errors of 21% and 29%, and only about 2% if one does not. As these two major differences in values appear to be of little or no consequence to the design, we feel justified in adopting the 2% figure. All other errors are between 0% and 8%. It should be noted that these are not errors in any real sense, just reflections of some differences in strategy or design knowledge, as described above.

6.2.4 Plan Selection

Plan selection is one place where the system adopts a simplified approach instead of using the full theory. Currently the plans are selected from the list in the order given, one at a time until the list is exhausted. If the plans are reasonably ordered this will work quite acceptably. This is adequate as each specialist appeared only to require one or two plans. We suspect that with more debriefing of the designer for a wider range of requirements more plans will be discovered.

It may be that plan selection really isn't that much of a problem as it immediately appears. Due to the decomposition into subproblems in order to manage complexity, we expect few plans per specialist. If there are few plans to choose from then selection need not be very complex.

Suppose that there are two plans in every specialist in a fully specified AIR-CYL system. Given the current specialist hierarchy, there would be about 130 different sequences of plans that can be followed in order to achieve the design. A very large number of actual designs can arise depending on the actual values chosen. So even with a small hierarchy and a small number of plans at each point there are still a very large number of designs captured. In all probability the designer will gradually form preferences and will do very little work during plan selection.

6.2.5 Failure Handlers

The major omission here is that for many of the possible system failures there are no failure handlers provided. This is merely a matter of inserting more knowledge expressed in DSPL; i.e., more FHs of more or less the same content as those already written. Everything has been provided so that they could be written.

The FHs for the major error conditions are all provided, and are demonstrated in Appendices B and C. Other cases besides those shown have been tested. Not all of the User FHs have been written either, but these are taken care of adequately by the default failure handlers. Although the default FHs are not as sensitive to minor variations they provide enough capability to do system testing and even full design, provided that no unusual errors occur.

Failure handlers are provided for the major failures that can occur in steps and tasks, but nothing has been provided for specialists. This does not mean that the system will not work; in fact, every effort has been made to ensure that the system will always continue, even with missing knowledge. Some minor modifications to the DSPL interpreter need to be made to make the calls to the two FHs that a specialist will need.

6.2.6 Redesigners

In the current implementation the redesign processing is carried out by functions embedded in the interpreter, and the user has no control over what is done or how it is done. Step redesign and task redesign are invoked from failure handlers using (STEP-REDESIGN-WITH-SUGGESTIONS) and (TASK-REDESIGN-WITH-SUGGESTIONS) respectively. Our goal is to extend DSPL to allow the user to have some control over the redesign action. This is especially needed for redesign in specialists, but would also be useful for tasks as there are several possible strategies and substrategies.

Currently redesign has only been implemented for steps and tasks. We feel that this has allowed adequate testing of the theory and its implementation at this time. Refinement of the theory of specialist redesign and its subsequent implementation will have high priority for our continuing research.

We expect that many of the roughly one hundred attributes will have some redesign knowledge associated with their steps, however only a few have actually been encoded in DSPL. We feel that enough redesigners have been written to show the viability of the failure handling theory.

Step redesigners have been written to include INCREASE and DECREASE suggestions; however, no CHANGE suggestions can be handled at this time. This needs further research, although it is clear that the change section of a redesigner will function first as a decision maker, to decide in which direction and, if necessary, how much to make the change.

6.3 The DSPL System

6.3.1 System Setup

The system setup phase is responsible for completely initializing the DSPL system with all the Lisp functions required and all the design knowledge required. Once everything is loaded and indexed, and some checking has taken place, the design proper can begin.

Each declaration takes the information provided and indexes it. Thus all agents are accessible via their name. Constraints have their syntax checked to make sure that the TEST is acceptable, but apart from that there is little syntax checking during agent declarations. Tasks and steps note the names and types of failure handlers used, so that it need not be done at run-time. In addition, the steps add their names to some information kept with their controlling task. The task accumulates a list of the steps used, a list of the attributes controlled, and an AttributeToStepTable, to be discussed later.

Another part of the initialization reads in a table of standard fractions are expressed as decimals (e.g., 1/5 expressed as 0.2000). This table is used to "standardize" values. This will be discussed below.

6.3.2 Attribute Tables

The attribute table for a task, known in the implementation as an AttributeToStepTable, allows a task to associate a particular attribute name with a step over which it has control. That is, given an attribute name; from a suggestion, for example; a task can find out if there is a step over which it has control that is responsible for that attribute. This is essential during redesign, as suggestions name attributes, and the task's redesign strategy must select steps to use during the redesign. The table is organized as on A-list of the form ((AttributeName

StepName)....). The table also allows a task to determine whether a suggestion is not for it and should be passed upward. Attribute tables will be needed by Specialists too, but they have not been implemented.

6.3.3 Standard Measures

In design, it is important to try to produce standard sizes in many places during the design. Making the width of a block 0.5 inches would mean that it could be cut from standard bar stock, whereas a width of 0.4917 would involve expensive machining, or a special order. Consequently DSPL allows requests to STANDARDIZE-UP, STANDARDIZE-DOWN and STANDARDIZE-NEAREST. So, for example, a value of 2.4936 can be made into 2.5 or 2.4844 (i.e., 31/64ths), by standardizing up or down, respectively. Standardizing to the nearest value would give 2.5 as the value.

This is implemented by using a binary search in a table of standard fractions expressed as decimals. The value in question is stripped of its integer part prior to the search, and the resulting decimal part is added back on again afterwards.

6.3.4 Tolerances

The arithmetic operators +, -, /, *, and ↑, as well as the relational operators =, ~=, >, >=, <, and <= are all redefined from their standard definitions in order to work with tolerances. Lengths with tolerances are expressed using (LNGTH <value> <+ve.tolerance> <-ve.tolerance>). It is also possible to express whether the value is accurate to one, two or three decimal places, for example, (LNGTH 0.123 0.01 0.0 'ThreeDP). Various functions are available to access the individual parts of a LNGTH, as well as to provide useful values such as the minimum value, the maximum value, the largest tolerance, and the total tolerance.

6.3.5 Messages

Messages are generated by the system. If the system fails at any point the failure messages will be displayed; only then does the user need to know anything about them. The contents of messages have already been described.

If the system trace is turned on then the flow of message passing will be displayed. The things that can be traced are KNOWNS, DECISION, Selector, Sponsor, Specialist, Plan, Task, Step, Redesigner, KB-FETCH, KB-STORE, and TEST-CONSTRAINTS.

6.3.6 The DSPL Interpreter

DSPL is interpreted by a set of functions that are highly structured to reflect the structure of the language. For example, there are parts of the interpreter responsible for executing steps, step bodies, and step body items. Each part of the interpreter can be passed a message, and will return one, even if it is only the message received from below. Each part is responsible for detecting failure messages and acting appropriately. In situations where a failure needs to be handled a system failure handler appropriate for the failing situation is called. Some failure handlers call for redesign, in which case the appropriate redesigner is activated.

Some care has been taken to make the variables used in KNOWNS sections and in DECISION assignments entirely local to that agent. Those variables will be dynamic and will only be active while the agent is active. This means that the same variables can be used in many places without problems occurring. It also allows embedded constraints to use the variables currently active, instead of, for example, repeating all the calculations of the superior agent.

Throughout the interpreter strange situations will always produce a failure -- either "Use of plan language problem" if there is some strange use of DSPL, or "Programming problem" for any unexplained and unexpected error condition.

6.4 The Design Data-base

6.4.1 The Frame Hierarchy

The design is recorded in frames constructed using a version of FRL [Roberts 77]. The data-base has been kept simple, and there is no record of component/subcomponent relationships. The frame for each component part consists of an entry for each attribute associated with that part. Figure 34 shows the frames in the DDB.

As the design proceeds an instance is created for each part as soon as it is needed. The final state of the instances is the design; that is, values for all of the attributes in the design. The user need not know about the instances and will use the part name and the attribute name in a FETCH or STORE; e.g., (KB-STORE 'Head 'HeadWidth xyz). The hierarchy can be used to specify defaults for things that really don't need to be designed but which are included for completeness, and can also be used to house constraints to be triggered on the addition of new values. A frame is provided to store the descriptions of the attributes in the set of requirements. These

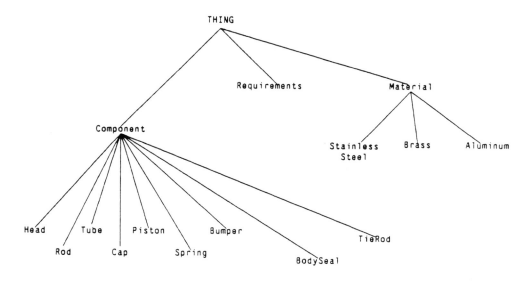

Figure 34: The Frames of the Design Data-Base

are used during interaction with the user when the requirements are being input. Default requirements are also available should they be needed.

6.4.2 Changes and Updates

During implementation "alterations" were referred to as "changes". In addition it was decided that to simplify things slightly one revision would constitute an update. Consequently an update of the drawing is done after every task. The drawing is merely the current state of the instances in the design data-base. Most of the complexity arises due to the need to be able to backup and undo changes and updates after failure.

6.4.2.1 Changes

Every use of KB-STORE in an agent results in a change. The changes are stored in a list called the Collected Changes with the most recent at the front. Each change is a record structure that includes the frame name and slot name, the old value (if there is one), the new value, the name of the agent making the change and its type, the time, the ID of the drawing to be affected. A stack is used to note the current state of the changes by placing a pointer to the current front of the list on the stack

117

wherever such a note is needed. Thus it is possible to revert to a prior design state within a task by noting the current state prior to use of a step or redesign, and then to "throw away" every change that is due to the failing action.

As each store should be able to trigger constraints the frame is accessed for every change made and a "dummy" store is made in order to see if any constraints trigger. This dummy store does not alter the state of the instance. If the change produces constraint failure the failure message is returned for analysis by the agent doing the change.

The KB-FETCH is used to get a value from the design data-base. It needs to inspect the collected changes first to see if a value is stored there, otherwise it looks in the appropriate instance. The fetch will retrieve only the most recent value for that attribute from the collected changes. It can return a value, or a failure message indicating that the slot was empty or that some other problem occurred.

6.4.2.2 Updates

An update will get done on exit from a successful task. An update is recorded in a record structure containing its ID, the name and type of the agent making the update, the time, the ID of the drawing being updated, the ID of the resulting drawing, and the collected changes. Updates are recorded on a list with the most recent first. In addition a history of updates is kept that records whether the update is being "done" or "undone". The update is "done" by adding it to the front of the update list, incrementing the drawing ID, and by making all the alterations to the instances in the design data-base as described by the collected changes. As these changes have all already triggered any relevant constraints this need not be done again. After use the collected changes are discarded from the list of changes as they are still recorded in the update.

On entry to a plan the state of the design is kept by remembering the current drawing ID on a stack. A stack is used as other plans may be activated in specialists mentioned in this plan. On successful exit from a plan this note of the entry situation is discarded. A failing exit from a plan needs to undo any updates that have been made until the drawing ID is the same as it was on entry to the plan. As each update is available and every update includes its changes and every change includes the old value, this process is possible.

6.4.3 Performance

Memory and timing information for the system are given in Figures 35 and 36. These timings are for a system with no compiled Lisp code, and are on a DECsystem-20.

```
*What*             *36 bit words*

LISP + FRL              118272
+ Fns. for DSPL         177152
with DSPL for App. D    190976
with DSPL for AIR-CYL   217600
```

Figure 35: Memory Utilization

```
*What*            *Time*
```

AIR-CYL with no failure
 and full system trace.
  .... CPU   54902 msec
  .. Garbage Collecting   6281 msec
  .... Clock   106441 msec

AIR-CYL with failure
 and full system trace.
  .... CPU   73671 msec
  .. Garbage Collecting   7879 msec
  .... Clock   200565 msec

AIR-CYL with no failure
 and no system trace.
  .... CPU   45467 msec
  .. Garbage Collecting   7539 msec
  .... Clock   211543 msec

AIR-CYL with failure
 and no system trace.
  .... CPU   59613 msec
  .. Garbage Collecting   7505 msec
  .... Clock   234826 msec

With 1.8 "average" load factor,
 but load varying from 1 to 2.5
With all input typed quickly and promptly.

Figure 36: Timing Figures

6.5 Summary

We have presented the Air-Cylinder domain, the AIR-CYL system itself, the major features of the DSPL system as it has been implemented, and a description of the Design Data-Base.

In the next chapter we will present a research agenda for the study of design problem-solving.

7 Design Problem Solving: A Research Agenda

Many detailed suggestions for further research have been given in [Brown 84]. We will present some of the main ideas here. These will be divided into Improvements to the DSPL System, Problem-solving in DSPL, and Other Research directions. These suggestions move from specific suggestions about support tools for building DSPL-based systems, to improvements to the problem-solving power of DSPL, to general suggestions about what topics should be pursued by AI in Design researchers.

7.1 Improvements to DSPL System Support

There are a variety of ways of making it easier to build DSPL-based systems.

7.1.1 Interfaces

7.1.1.1 The Design Language

DSPL acts as an interface with the knowledge engineer, and, consequently its syntax and semantics are of great importance. The representations adopted for the different types of knowledge in routine design must be easy to use, and general and expressive enough to be of use across domains.

DSPL is being tested on other design problems in different domains in order to discover its utility. Discovered flaws will lead to improvements in the language.

Further research will be needed to investigate language forms for DSPL-like languages that are most easily understandable by designers themselves. At present we are not concerned with designers being able to write in the language directly as we will assume the assistance of a knowledge-engineer. Various aids to expert system construction would be useful and will be discussed in the next section.

7.1.1.2 Input and Output

In order to make this system available for anyone other than the implementor the system's interaction with the user must be improved. In addition it should be possible for the system to display more about the state of the design and the problem-solving during execution. Graphical output of the design on completion, or even during execution, would improve the system considerably.

7.1.1.3 Explanation

Many people argue that an expert knowledge-based problem-solving system isn't an "expert system" unless it has an explanation capability. While we do not necessarily agree with this point of view such a capability significantly enhances a system.

Initial research has been completed at both Ohio State and at WPI to build explanation onto DSPL [Kassatly 87] [Chandrasekaran 88b]. At Ohio State this work is being continued to investigate a full range of explanation methods for design. Given the task-level structure and expressiveness of DSPL we expect the resulting explanations to be of higher quality that those of rule-based systems.

7.1.2 Design System Builder's Aids

The Xerox Lisp machine version of DSPL includes a syntax checking editor for DSPL. It ensures correct input to the system and provides skeletal forms of DSPL expressions for the user to fill in. This speeds up the input of knowledge and provides consistent formatting throughout the knowledge gathering sessions.

That version of DSPL also includes indexing of agents, as well as structured display of the agents in the system. Use of the DSPL system could be greatly enhanced by the development of a more complete set of system building support tools, such as those presented below.

7.1.2.1 Design Language Help System

An on-line help facility should be available so that the semantics of the language are explained to the user. Examples of the use of the language could be given to the user.

7.1.2.2 Design Language Knowledge Gatherer

This tool would be a first-cut knowledge acquisition tool. It would not know about the domain, as a full blown acquisition system would need to, but instead would know about DSPL and its use. It would guide the user through a session of writing DSPL, keeping track of things already written and things yet to be expressed. It would prompt the user whenever he or she doesn't know what to specify next. An editor, indexer, structure display and help system could all be used during the gathering process to keep the user informed.

A prototype Knowledge Gatherer has been implemented at WPI on a VAX in VAXlisp [Chiang 87]. It includes strategies that guide the prompting of the user by the system. It has a full knowledge of DSPL syntax. Exactly in which order the user prefers to present the design knowledge is being investigated. A more general discussion about the role of generic tasks in knowledge acquisition can be found in [Bylander 87].

7.2 Problem-solving in DSPL

While investigating design problem solving we are trying very hard to extend the language and the theory only when an application clearly demands it. Our methodology is to use the simplest approach until it can be shown that it is inadequate and needs to be extended, and then extend it. This approach can be, and has been, criticized as being restrictive. However, it tends to prevent us from including anything superfluous, thus producing a much tighter match between the expert's knowledge and the language, and consequently a much stronger theory.

There are, however, a number of research topics that grow directly out of the development of DSPL and the AIR-CYL system. We present these below. Note that several of them are of great general importance for the AI in Design community.

7.2.1 Conjunctive Suggestions

In simple situations suggestions about what might be altered in order to remedy the difficulty will often be in the form "Suggestion1 or Suggestion2 or Suggestion3 etc". Each suggestion is considered to be independent of the others and each will be considered until one is found to work. Consider the situation where the form of the failing constraint is such that no one value can be changed to satisfy it but two changes might. More research is needed to find out how this can be accomplished.

7.2.2 Plan Selection

The issue of how plans get selected is far from resolved. While it is clear that many pieces of information are involved we do not yet know how they are combined during selection.

The selection mechanism adopted in DSPL can express many different plan selection strategies. We need to look for patterns in the methods of sponsorship and selection, and then restrict the language to just those expressions that are needed in the language of selection. At present the language allows too much.

Conversely, there are several things that need to be added to DSPL to assist with plan selection -- the language doesn't deal fully with requirements; more information about how the last plan failed should be available; the structures of plans should be comparable; dependencies should be represented and examinable; dependency measures should be examinable. However, we are confident that all of these can be achieved by extending the currently available functions.

7.2.3 Agent Memory

We consider each specialist to have some "memory" of prior use, but exactly how much detail is retained is not entirely clear. However, it is not reasonable to expect the specialist to always recognize that a situation is exactly the same as some prior situation. Notice that this implies that some kind of history of failure be kept, and that it is accessible to the agent. However, it need not (and should not) be global for all the design agents to access.

We have already included in the theory the need for a specialist to remember the plan it executed to achieve success. This is needed to handle redesign requests. Some kind of local memory would also be of use during explanation.

7.2.4 Multiple Designs

It appears that sometimes during routine detailed design a designer will produce more than one alternative detailed design of some subportion and then evaluate their consequences for later design decisions. This needs to be examined more closely and the knowledge and control structures determined.

In one domain being investigated at WPI the designer produces three alternatives for a small portion of the design and then extends each to a full design, one after the other, with comparisons between the results. We are still analyzing this domain, but it is clear that DSPL will have to be changed in order to be able to capture this design activity and the knowledge involved. It does not appear that the designer considers this as three independently parallel designs.

7.2.5 Failure-handling Strategies

The strategies that control backup at the task and specialist level have only been outlined quite roughly so far. Some attempt needs to be made to find out which strategies are preferred by designers in which situations.

In addition, exactly how to use those suggestions that have been passed up but not yet used needs to be investigated. Currently we believe that many of the suggestions that are passed up untried are so costly that they can be ignored, but that need not be true in general. The specialist needs more local knowledge to make these kinds of decisions.

The issue of exactly how suggestions are obtained has been finessed in the current theory. We need to investigate what knowledge and analysis is needed in order to produce a suggestion. That research has already begun at WPI.

7.2.6 Evaluating the Design

One aspect of the design process that we observed but have ignored in the implementation is that of evaluating the design. The air-cylinder design concluded with some cost and weight evaluations and some subsequent modifications to the design. As these modifications were not making substantial changes to the design they were ignored for the AIR-CYL implementation.

In general, as design proceeds many pieces of critical knowledge about (for example) weight, cost, manufacturability, and design standards may be bought into play, either directly by the designer (e.g., "I wonder how easy this is to make?") or by associative processes over which the designer has little or no control.

In addition, use of mental or qualitative simulation can play a part, or the designer can use computer-based numerical analysis tools, such as simulation, or stress analysis methods, as was pointed out in Chapter 1. Even if computer-based tools are used, there is still knowledge involved, such as which analysis method to use, or how to interpret the results of the analysis.

7.2.7 Post-design Re-design

If post-design evaluation of the design leads to dissatisfaction then the design cannot be considered to be complete, and more design activity remains. It remains to be seen whether DSPL's re-design/redesign process will suffice here, or whether some other approach is needed.

7.2.8 Rough Design

Currently it is assumed that all rough design is done at the start of the design activity in order to establish that design proper is worth pursuing. There is no need for all the rough design to be done initially. Some could be done at the start of a major subcomponent in order to quickly establish in a sort of "back-of-the-envelope" manner whether design is likely to succeed. Breau [Breau 85] has found an example of this in a packaging design situation.

As the current implementation allows rough-design to proceed under the control of any plan, rough design can actually be done at any point during the design activity. Notice that even if the flow of control is top-down the values may be chosen in ways that appear to be bottom-up, inside-out, top-down or mixed.

There is one aspect of rough-design that we have not yet tried to capture and that is the "rough" value. For example, when designing we often start by assigning an attribute a small range of values -- "it's between 3 and 4 inches". Values in a design system may need to have qualities of "precision", "accuracy", and "confidence". Accuracy is captured at present by using a form that includes a value and its tolerances -- (LNGTH 2.3 0.001 0.02). Precision has to do with how precisely one is able to state the range of possible values -- less than 4. Confidence has to do with how allowable it is to alter that value -- "a good starting value for this is 3 inches + or - 1/10th, but if you want you can change it if it doesn't look right". Clearly these all interact in various ways, and will pose problems for the system -- what's "roughly 3 but less than 4" plus (LNGTH 2.3 0.001 00.02)?

There are distinct differences between specification checking, rough design, design and redesign activity. We feel that the differences between design and redesign have been shown fairly well, but we have so far failed to identify all the differences between design and rough-design. Thus there is a need to investigate these differences.

7.2.9 Relaxation of Requirements

One possible way to deal with failures is to attempt to relax one or more of the requirements. Clearly some requirements can be "softer" than others. If a lot of effort has already been expended on a design by both machine and human then relaxation is appropriate. It may be possible for the system to choose requirements to relax, but a lot of special knowledge would be necessary to implement that. Even knowing when to ask for a relaxation will be difficult. Relaxation of requirements is a matter for future research.

7.2.10 Automatic Construction

Another interesting possibility for the system is the automatic ordering of steps and tasks. At present one has to be aware of the exact order that the designer uses so as to be sure that all required values are available before each agent acts. Due to the commonly observed difficulty in getting one's expert to recall exactly what he or she does, it will often be possible to get only a partial ordering. In fact, provided the dependencies are immediately available or can be deduced, the agents could be presented to the system for it to decide on a reasonable order. This could be done if the designer is is doubt or doesn't care. See [Kumar 85] for one approach to this idea.

It may be possible to establish from dependency information those mutual dependencies that need to be broken by rough design, and also those values that have very high dependency measures and should therefore be chosen during rough-design to ensure that the design is possible.

We need to more firmly establish exactly what the principles are for deciding which pieces of knowledge correspond to specialists. Of course, how the designer presents and uses the information plays a large part, as the hierarchy must match the designer's conceptual view of the problem, but we suspect that there are underlying rules or heuristics that lead to the designer's conceptualization:

- the dependencies must be observed and incorporated into the structure;
- common information and decisions should be raised to their highest relevant level;
- easier decisions that are useful for the rest of the design, if they can be made, get made first, but often the easy decisions don't contribute much, and often the hard bits of design get done first so that a failure there will quickly lead to complete failure before too much energy has been expended;

there is some relationship to the structure of the thing being designed;

• there is some relationship to the functions of the thing being designed.

There is much to be done here, as there is in the study of structures for diagnosis, to establish the principles of hierarchy construction.

7.2.11 Performance Degradation

We suspect that as designs get to be on the edge of being Class 3 the performance of the system will degrade in interesting ways. By observing this behavior, and the ease with which design knowledge for an object can be captured, we hope to be able to make observations about subclasses of design more precise than the classes that we have already outlined.

If there are many dependencies between attributes then the system can be expected to fail more often. If there are many mutual dependencies then the system will be making a lot of early decisions that may be based on limited information and may well be wrong, leading to poor results from the design phase. If the dependencies do not allow the design knowledge to fall into easily divisible groups then it will be very difficult to produce a design system at all. Many dependencies will cause the effects of a failure to be much stronger as many more alterations will have to be made after a suggestion has been followed.

7.2.12 Use of Defaults and Catalogs

In order to make the system really usable we have to look at how designers use catalogs during the design process. Having decided on a particular type of component their next decision may well depend on exactly what is in a catalog. Certainly this is one way of obtaining the size of a component. It is likely that some values are not calculated but instead they are retrieved as default values from knowledge about components. This has yet to be investigated in the AIR-CYL context, even though it was used in the prototype system [Brown 83].

7.2.13 Evaluation of the System

It is always difficult to evaluate the performance of an expert system, and evaluation of most expert systems is usually inadequate. Unfortunately, AIR-CYL is no exception. Not only will designers differ in their designs, as there is usually no single answer, but even if the figures are different from the "standard" the design may still be acceptable or even excellent. As pointed out earlier, slight deviation from an expected value will affect many other values, and provided they are still mutually consistent the resulting design may be just as good.

The results from the AIR-CYL system produce a design that varies only slightly from the designer's, and in at least one way is superior when judged by the designer's own heuristics. What is being measured is not really the performance of the system, but rather how well the knowledge has been captured in DSPL.

7.3 Limitations of Class 3 Approach

Of course, we are quite aware that there are bound to be other examples of routine design tasks that cannot be brought under the hierarchical design plan selection paradigm in a natural way. Even if it is true in principle that design is a process of choosing and refining plans, our ability to write expert systems for design is very much a function of the generic classes of plans that we are able to describe and manipulate.

We would like, as a result of further research, to be able to characterize the kinds of design problems for which the plan refinement approach will lead to effective expert systems. This will only come with further research into problems that would prevent adequate performance of a design expert system, and with experience with other problems in the domain.

There are a number of different domains, besides the air cylinder domain, that are being investigated. At Ohio State DSPL is being used to examine the connections between routine design and routine planning. A prototype mission planning system has been built that designs flight plans for aircraft [Herman 86]. Another application, in the domain of chemical engineering, designs distillation columns from specifications of the components of the feed to the column and certain parameters describing the columns environment.

At WPI we are studying a Package design domain, and have just started on gear design and electrical circuit design. Another application is concerned with the integration of constructibility knowledge with knowledge about the preliminary design of buildings.

7.4 Directions for Design Research

Given the many different types of knowledge and problem-solving involved in the different types of design activity there are many different directions for design research. We believe that incremental exploration from routine design towards creative design is the best approach. This will allow discovery of most of the ingredients of a more complex form of design prior to investigating it.

It is clear that research into design -- be it at the level of protocol analysis of human designers, or at the level of building knowledge-based computational models that depend on hypotheses about the ingredient generic tasks -- can only be compared, contrasted and evaluated if we have a way of classifying design tasks. A particular human design activity will vary depending on the expertise of the person, the domain, the level of detail of the design problem as determined by the requirements and the desired output, whether it is design or redesign, and other factors. If these axes can be established, a design task can be classified by determining its position in the resulting N-dimensional space.

A theory of design activity belonging to a particular classification can only be fully explained in terms of the types of knowledge and associated control strategies that are used. Thus, different combinations of generic tasks will correspond to different coordinates in that space. Observable changes in the design protocols, such as "phases" of design, can be explained in terms of underlying knowledge and control. Our Classes 1, 2 and 3, are clearly only a starting point for such an analysis.

7.4.1 Investigating Compilation of Design Knowledge

We are convinced that one of the research directions with high payoff is the study of how all forms of design knowledge become compiled. The compilation of deep structural and functional knowledge [Sembugamoorthy 86] should provide a skeleton on which to hang other types of knowledge. An interesting and highly relevant piece of research can be found in [Araya 87].

Issues that will have to be addressed include the indexing, retrieval and specialization of past designs, design plans and object knowledge, along with their abstractions; the use of analogy to produce new design knowledge from old [Carbonell 83]; the debugging and patching of partially correct design plans [Sussman 73]; how to reason with functional representations in order to build a functional description of the object that will satisfy the design requirements [Freeman 71, Bowen 85]; the adjustment of design knowledge as it becomes more routine [Brown 87].

7.4.2 Failure Analysis

The need for a better theory of the analysis and handling of failure during design has already been discussed above. For highly compiled knowledge about routine design, and for the most detailed stages of design, failure handling is fairly easy to describe. However, as design becomes less routine there are many open problems concerning failure recognition, credit (blame) assignment, and the types of knowledge and control used during redesign.

The difficulty of this area is compounded by the fact that there needs to be a different type of failure handling depending on what type of problem-solving is being used (e.g., failure handling during Generate and Test, as opposed to Decomposition plus Constraint Posting) (see Chapter 2).

7.4.3 Decomposition

Another important research topic is the study of how designers decompose design problems in order to handle complexity. With well known types of objects, in designs which have routine subproblems, the decomposition may be known or can be derived by analogy. For more innovative designs analogy may be possible, from existing designs or existing design plans.

Decomposition may require a functional understanding of the object that could satisfy the requirements, which could then lead to hypotheses about possible structure, and hence a decomposition. Whatever the method of decomposition, we will need to understand the knowledge about how to compose the solutions to the subproblems produced by decomposition.

7.4.4 Design by Weak Methods

Observation of designers, by researchers at Rutgers University, the University of Massachusetts, and ourselves, suggests that for some routine design problems the "decompose and decide" approach works to the point of choosing components, but, due to the high degree of dependency between the resulting parameters, another method needs to be used in order to determine the values of those parameters.

This method is a form of hypothesis and test problem solving, where initial values are chosen for the attributes in an heuristic way, the design is evaluated, and, if unacceptable, some parameters are chosen to be adjusted. Knowledge and experience are used to solve both the credit (blame) assignment problem (i.e., the choice of parameters to change), as well as to determine by how much they should be changed.

This evaluate, blame, adjust cycle is repeated until the designer is satisfied with the design. This approach can be, but need not be, constrained to be a hill-climbing problem.

The Air Cylinder designer we observed used something like this method to "tweek" the design slightly, by propagating small changes in dimensions backwards and forwards along the longitudinal axis of the air-cylinder until the desired state was reached. The changes were very small and didn't result in the design being invalidated. These changes appear not to based on the same conceptual organization of the problem, but might be done by using spatial reasoning on a spatial representation of the component. This whole area of re-design after the design is complete needs investigation.

7.5 Conclusions

We have learned much during the last few years as a result of our efforts implementing systems, such as the AIR-CYL expert system, and from ongoing research. These implementation efforts are not complete, but we now have a better understanding of the strengths of and weaknesses of the approach.

Our research has outlined a theory of routine design problem-solving, by identifying the types of knowledge used and the interactions between those types of knowledge. We have also shown the importance of rough design, of plan selection and of failure handling, and have at least a basic understanding of how these function.

The Design Specialists and Plans Language, DSPL, allows knowledge to be expressed in a problem-solving type dependent but design domain independent manner. An interpretive system that executes DSPL has allowed us to show the practicality of the approach by building the various expert systems.

While the idea of design detailing by hierarchical plan selection captures the essence of routine design problem solving, we know that there are many important aspects of design problem-solving and design expert system building that need continued investigation.

We have presented a collection of research problems, ranging from the development of tools to aid in the building of DSPL-based design expert systems, to extensions to DSPL's knowledge and problem-solving. We have also discussed some more general problems that we feel are appropriate directions for the next round of research into AI in Design.

Much research remains to be done before we fully understand design problem-solving and how we can build adequate systems to do it. However, we feel that by using a hierarchically structured system of conceptual specialists with plan selection and failure handling we have captured the essential qualities of routine design. This approach has led us to discover many interesting and difficult issues, and has suggested many new areas for research. New applications are being suggested daily. There is no doubt that the current excitement about AI in Design is completely justified.

References

[Araya 87] Araya, A. and S. Mittal.
Compiling Design Plans from Descriptions of Artifacts and Problem Solving Heuristics.
In *Proceedings of the 10th International Joint Conference on Artificial Intelligence*, pages 552-558. Milan, Italy, August, 1987.

[Balzer 81] Balzer, R.
Transformation Implementation: an Example.
IEEE Trans. Software Engineering SE-7:3-14, 1981.

[Barstow 84] Barstow, D.
A Perspective on Automatic Programming.
The AI Magazine 5(5), 1984.

[Bowen 85] Bowen, J.
Automated Configuration Using A Functional Reasoning Approach.
In *Proceedings of the Society for the Study of AI and Simulation of Behavior*, pages 1-14. University of Warwick, England, 1985.

[Breau 85] Breau, R. and D. C. Brown.
A Study of Class 3 Design Problem Solving.
In *Second Annual Northeast Regional Conference*. ACM, October, 1985.

[Brown 83] Brown, D. C. and B. Chandrasekaran.
An approach to expert systems for mechanical design.
In *Trends and Applications '83*, pages 173-180. IEEE Computer Society, NBS, Gaithersburg, MD, May, 1983.

[Brown 84] Brown, D. C.
Expert Systems for Design Problem-Solving using Design Refinement with Plan Selection and Redesign.
PhD thesis, The Ohio State University, August, 1984.

[Brown 86] Brown, D. C. and R. Breau.
Types of Constraints in Routine Design Problem-Solving.
Applications of AI in Engineering Problems.
Springer-Verlag, 1986, pages 383-390.

[Brown 87] Brown, D. C., and W. N. Sloan.
Compilation of Design Knowledge for Routine Design Expert Systems: An initial view.
In *Proceedings of ASME International Computers in Engineering Conference, New York*, pages 131-136. 1987.

[Bylander 83] Bylander, T., S. Mittal and B. Chandrasekaran.
CSRL: A language for expert systems for diagnosis.
In *Proceedings of the 8th International Joint Conference on Artificial Intelligence*, pages 218-221. Karlsruhe, W. Germany, August, 1983.

[Bylander 86] Bylander, T., and S. Mittal.
CSRL: A Language for Classificatory Problem Solving and Uncertainty Handling.
AI Magazine 7(3):66-77, 1986.

[Bylander 87] Bylander, T. and B. Chandrasekaran.
Generic Tasks for Knowledge-Based Reasoning: the "Right" Level of Abstraction for Knowledge Acquisition.
International Journal of Man-Machine Studies 26:231-243, 1987.

[Carbonell 83] Carbonell, J. G.
Learning by Analogy: Formulating and Generalizing Plans from Past Experience.
Machine Learning.
Tioga Press, 1983, pages 137-60.

[Chandrasekaran 80]
Chandrasekaran, B., S. Mittal and J. W. Smith.
RADEX - Towards a computer-based radiology consultant.
Pattern Recognition in Practice.
North Holland Pub. Co., Amsterdam, Holland, 1980, pages 463-474.

[Chandrasekaran 81]
Chandrasekaran, B.
Natural and social system metaphors for distributed problem solving: Introduction to the issue.
IEEE Transactions on Systems, Man, and Cybernetics SMC-11(1):1-5, January, 1981.

[Chandrasekaran 83a]
Chandrasekaran, B.
Towards a Taxonomy of Problem-Solving Types.
AI Magazine 4(1):9-17, 1983.

[Chandrasekaran 83b]
    Chandrasekaran, B. and S. Mittal.
    Conceptual Representation of Medical Knowledge for Diagnosis by Computer: MDX and Related Systems.
    Advances in Computers.
    Academic Press, 1983, pages 217-293.

[Chandrasekaran 86]
    Chandrasekaran, B.
    Generic Tasks in Knowledge-Based Reasoning: High-Level Building Blocks for Expert System Design.
    IEEE Expert 1(3):23-30, 1986.

[Chandrasekaran 87]
    Chandrasekaran, B.
    Towards a Functional Architecture for Intelligence Based on Generic Information Processing Tasks.
    In *Proceedings of the International Joint Conference on Artificial Intelligence.* IJCAI, 1987.

[Chandrasekaran 88a]
    Chandrasekaran, B.
    What Kind of Information Processing is Intelligence? A Perspective on AI Paradigms, and a Proposal.
    In Partridge, D. and Wilks, Y. (editors), *Foundations of AI: A Sourcebook.* Cambridge University Press, 1988.

[Chandrasekaran 88b]
    Chandrasekaran, B., M. C. Tanner and J. R. Josephson.
    Explanation: the Role of Control Strategies and Deep Models.
    Expert Systems: The User Interface.
    Ablex Publishing Corporation, Norwood, New Jersey 07648, 1988, pages 219-247.

[Chiang 87]
    Chiang, T. Y-L. and D. C. Brown.
    DSPL Acquirer: A System for the Acquisition of Routine Design Knowledge.
    In *Proceedings of The 2nd International Conference on the Applications of Artificial Intelligence in Engineering*, pages 95-110. Computational Mechanics Publications, Boston, MA, August, 1987.

[Clancey 84]
    Clancey, W.
    Classification Problem Solving.
    In *Proceedings of the 4th National Conference on AI*, pages 49-55. 1984.

[Dixon 84]
    Dixon, J. R., M. K. Simmons and P. R. Cohen.
    An Architecture for Application of Artificial Intelligence to Design.
    In *Proceedings of the 21st Design Automation Conference*, pages 634-640. IEEE, 1984.

[Doyle 80]   Doyle, J. and P. London.
               A selected descriptor-indexed bibliography to the literature on
                   Belief Revision.
               ACM SIGART (71):7, April, 1980.

[Duda 79]   Duda, R., J. Gaschnig and P. Hart.
               Model Design in the Prospector Consultation System for Mineral
                   Exploration.
               Expert Systems in the Microelectronic Age.
               Edinburg University Press, 1979, pages 153-167.

[Fahlman 73]   Fahlman, S. E.
               A planning system for robot construction tasks.
               AI-TR 283, MIT, AI Lab., MIT, Cambridge, MA, May, 1973.

[Frayman 87]   Frayman, F. and S. Mittal.
               COSSACK: A Constraints-Based Expert System for Configuration
                   Tasks.
               In Sriram, D. and Adey, R. A. (editors), *Knowledge Based Expert
                   Systems in Engineering: Planning and Design*, pages 143-166.
                   Computational Mechanics, August, 1987.

[Freeman 71]   Freeman, P. and A. Newell.
               A model for functional reasoning in design.
               In *Proceedings of the International Joint Conference on Artificial
                   Intelligence*, pages 621. IJCAI, 1971.

[Friedland 79] Friedland, P.
               Knowledge-based experimental design in molecular genetics.
               In *Proceedings of the 6th International Joint Conference in Artificial
                   Intelligence*, pages 285-287. IJCAI, Tokyo, Japan, 1979.

[Gini 83]   Gini, M. and G. Gini.
               Towards Automatic Error Recovery in Robot Programs.
               In *Proceedings of the International Joint Conference on Artificial
                   Intelligence*, pages 821-823. IJCAI, August, 1983.

[Goel 88]   Goel, Ashok.
               *Integration of Model-based and Case-based Reasoning for Design
                   Problem Solving.*
               PhD thesis, Department of Computer and Information Science, The
                   Ohio State University, Columbus, Ohio 43210, 1988.
               Forthcoming Ph.D dissertation.

[Goldstein 74] Goldstein, I.
               Summary of MYCROFT: a system for understanding simple picture
                   programs.
               Artificial Intelligence 6(3):249-288, 1974.

[Gomez 81] Gomez, F. and B. Chandrasekaran.
Knowledge Organization and Distribution for Medical Diagnosis.
IEEE Transactions on Systems, Man, and Cybernetics
SMC-11(1):34-42, January, 1981.

[Hayes 75] Hayes, P.J.
A representation for robot plans.
In *Proceedings of the International Joint Conference on Artificial Intelligence*, pages 181. IJCAI, 1975.

[Hayes-Roth 83] Hayes-Roth, F., D. Waterman and D. Lenat (editors).
Building Expert Systems.
Addison-Wesley, 1983.

[Herman 86] Herman, D., J. Josephson and R. Hartung.
Use of DSPL for the Design of A Mission Planning Assistant.
In *Proceedings of the Expert Systems in Government Symposium.*
IEEE, Washington D.C., October, 1986.

[Hewitt 72] Hewitt, C.
Description and theoretical analysis (using schemata) of PLANNER, a language for proving theorems and manipulating models in a robot.
TR 258, MIT, MIT AI Lab., Cambridge, MA, 1972.

[Johnson 85] Johnson, L. and E. Soloway.
PROUST: Knowledge-Based Program Understanding.
IEEE Trans. Softw. Eng. 11(3):267-275, 1985.

[Josephson 87] Josephson, J., B. Chandrasekaran, J. Smith and M. Tanner.
A Mechanism for Forming Composite Explanatory Hypotheses.
IEEE Trans. System, Man & Cybernetics :445-454, May/June, 1987.

[Kassatly 87] Kassatly, A. and D. C. Brown.
Explanation for Routine Design Problem Sovling.
In *Proceedings of The 2nd International Conference on the Applications of Artificial Intelligence in Engineering*, pages 225-239. Computational Mechanics Publications, Boston, MA, August, 1987.

[Kelly 82] Kelly, V. E. and L. I. Steinberg.
The CRITTER system: analyzing digital circuits by propagating behaviors and specifications.
In *Proceedings AAAI Conference*, pages 284. AAAI, August, 1982.

[Kelly 84] Kelly, Van E.
The CRITTER System: Automated Critiquing of Digital Circuit Designs .
In *Proceedings of the 21st Design Automation Conference*, pages 419-425. IEEE, 1984.

[Kumar 85]   Kumar, A., G. Kinzel and R. Singh.
A Preliminary Expert System for Mechanical Design.
In *Proceedings of the Computers in Engineering Conference*, pages 29-35. ASME, Boston, MA, August, 1985.

[Latombe 76]   Latombe, J-C.
Artificial Intelligence in Computer-aided Design : the TROPIC system.
Technical Report 125, Stanford Research Institute, February, 1976.

[Latombe 79]   Latombe, J-C.
Failure processing in a system for designing complex assemblies.
In *Proceedings of the International Joint Conference on Artificial Intelligence*, pages 508-515. IJCAI, 1979.

[Mackworth 77]   Mackworth, A. K.
Consistency in networks of relations.
Artificial Intelligence 8(1):99-118, 1977.

[Marcus 85]   Marcus, S., J. McDermott and T. Wang.
Knowledge Acquisition for Constructive Systems.
In *Proceedings of the International Joint Conference on Artificial Intelligence*, pages 637-639. IJCAI, 1985.

[Marcus 86]   Marcus, S. and J. McDermott.
SALT: A Knowledge Acquisition Tool for Propose-and-Revise Systems.
Technical Report, Carnegie Mellon University, Department of Computer and Information Science, 1986.

[McDermott 77]   McDermott, D. V.
Flexibility and efficiency in a computer program for designing circuits.
AI-TR 402, MIT, AI Lab., MIT, Cambridge, MA, June, 1977.

[McDermott 78]   McDermott, D. V.
Circuit Design as problem solving.
In Latombe, J-C (editor), *AI and Pattern Recognition in CAD*, pages 227-245. North-Holland, 1978.

[McDermott 80]   McDermott, J.
R1: an Expert in the Computer Systems Domain.
In *Proceedings of the 1st Annual National Conference on AI*, pages 269-271. AAAI, 1980.

[McDermott 82]   McDermott, J.
R1: a rule-based configurer of computer systems.
Artificial Intelligence 19(1):39-88, Sept, 1982.

[Miller 60]   Miller, G. A., E. Galanter and K. H. Pribram.
Plans and the structure of behavior.
Holt, Rinehart and Winston, 1960.

[Mitchell 83] Mitchell, T. M.
An Intelligent Aid for Circuit Redesign.
In *Proceedings of AAAI Conference*, pages 274-278. AAAI, 1983.

[Mitchell 85] Mitchell, T. M., L. Steinberg and J. Shulman.
A Knowledge-Based Approach to Design.
IEEE Trans. on Pattern Analysis and Machine Intelligence
PAMI-7(5):502-510, 1985.

[Mittal 80a] Mittal, Sanjay.
Design of A Distributed Medical Diagnosis and Data Base System.
PhD thesis, The Ohio State University, 1980.

[Mittal 80b] Mittal, S. and B. Chandrasekaran.
A conceptual representation of patient databases.
Journal of Medical Systems 4(2):169-185, 1980.

[Mittal 84] Mittal, S., B. Chandrasekaran and J. Sticklen.
PATREC: A Knowledge-directed Data Base for a Diagnostic Expert System.
IEEE Computer 17(9):51-58, 1984.

[Mittal 86] Mittal, S., C. Dym and M. Morjaria.
PRIDE: An Expert System for the Design of Paper Handling Systems.
IEEE Computer 19(7):102-114, 1986.

[Newell 80] Newell, A.
Reasoning, Problem Solving and Decision Process: The Problem Space as a Fundamental Category.
Attention and Performance, VIII.
L. Erlbaum, 1980, pages 693-718.

[Newell 87] Newell, A., J. E. Laird and P. S. Rosenbloom.
SOAR: An Architecture for General Intelligence.
Artificial Intelligence 33:1-64, 1987.

[Nilsson 73] Nilsson, N.J.
A hierarchical robot planning and execution system.
Technical Report 76, Stanford Research Institute, AI center, SRI, April, 1973.

[Punch 89] Punch, W. F.
A Diagnosis System Using a Task Integrated Problem Solver Architecture (TIPS), Including Causal Reasoning.
PhD thesis, Department of Computer and Information Science, The Ohio State University, Columbus, Ohio 43210, 1989.
Forthcoming Ph.D dissertation.

[Rich 81]   Rich, C.
                  A Formal Representation for Plans in the Programmer's
                      Apprentice.
                  In *Proceedings of the 7th International Joint Conference on Artificial
                      Intelligence* , pages 1044-1052. IJCAI, Vancouver, B.C.,
                      Canada, August, 1981.

[Roberts 77]   Roberts, R. B. and I. Goldstein.
                  The FRL Manual.
                  AI memo 409, MIT, AI Lab, Cambridge, MA, 1977.

[Sacerdoti 75]   Sacerdoti, E. D.
                  A Structure for Plans and Behavior.
                  Technical Report 109, AI Center, SRI, Menlo Park, CA, 1975.

[Schank 77]   Schank, R. C. and R. Abelson.
                  Scripts, Plans, Goals and Understanding.
                  Lawrence Erlbaum Associates, 1977.

[Sembugamoorthy 86]
                  Sembugamoorthy, V. and B. Chandrasekaran.
                  Functional Representation of Devices and Compilation of
                      Diagnostic Problem Solving Systems.
                  Experience, Memory and Reasoning.
                  Hillsdale, NJ: Erlbaum, 1986, pages 47-73.

[Shortliffe 76]   Shortliffe, E.
                  Computer-Based Medical Consultations: MYCIN.
                  Elsevier Scientific Publishing Co., 1976.

[Simon 81]   Simon, H. A.
                  The Sciences of the Artificial.
                  MIT Press, 1981.
                  1st Edition, 1969.

[Srinivas 78]   Srinivas, S.
                  Error recovery in robots through failure reason analysis.
                  In *Proceedings of the National Computer Conference*, pages
                      275-282. AFIPS, 1978.

[Stallman 77]   Stallman, R. and G. Sussman.
                  Forward reasoning and dependency-directed backtracking in a
                      system for computer-aided circuit analysis.
                  Artificial Intelligence 9:135-196, 1977.
                  also in MIT AI Lab memo. 380 -- 1976.

[Stefik 81]   Stefik, M.
                  Planning with Constraints.
                  Artificial Intelligence 16:111-140, 1981.

[Sticklen 83] Sticklen, J.
Manual for IDABLE -- Draft Version
AI Group, CIS Dept., OSU, 1983.

[Sticklen 87] Sticklen, J.
MDX2: An Integrated Medical Diagnostic System.
PhD thesis, Department of Computer and Information Science, The Ohio State University, Columbus, Ohio 43210, June, 1987.

[Sussman 71] Sussman, G. J., T. Winograd and E. Charniak.
MICRO-PLANNER reference manual.
AI Memo 203A, MIT, MIT AI Lab., Cambridge, MA, 1971.

[Sussman 72] Sussman, G. J. and D. V. McDermott.
CONNIVER reference manual.
AI Memo 259, MIT, MIT AI Lab., Cambridge, MA, 1972.

[Sussman 73] Sussman, G. J.
A computational model of skill acquisition.
AI-TR 297, MIT, AI Lab., MIT, Cambridge, MA, 1973.
Ph.D. dissertation, also in book, same title, Elsevier, 1975.

[Sussman 75] Sussman, G. J. and R. Stallman.
Heuristic techniques in Computer-aided circuit analysis.
IEEE Trans. on Circuits and Systems CAS-22(11):857, November, 1975.

[Sussman 77] Sussman, G. J.
Electrical Design -- a problem for AI research.
In *Proceedings of the International Joint Conference on Artificial Intelligence*, pages 894-900. IJCAI, 1977.

[van Melle 79] van Melle, W.
A Domain-independent Production-rule System for Consultation Programs.
In *Proceedings of the International Joint Conference on Artificial Intelligence*, pages 923-925. IJCAI, August, 1979.

[Wielinga 86] Wielinga, B. and J. Breuker.
Models of Expertise.
In *Seventh European Conference on Artificial Intelligence*. July, 1986.

Appendix A: Design Trace

This is a trace generated by the system. It has been edited for brevity and presentation in this format. The trace was only turned on for the Specialists. The trace is of a successful design with no selection of alternative plans. It shows the attributes designed by the system.

```
***** AIR-CYL Air-cylinder Design System *****
*** Version date: (4 26 84)
*** Todays date:  (5 4 84)
*** User: DCBROWN

*** Requirements input
The following options are available:
   1 --- to use a set of standard test/demo requirements
   2 --- to have requirements read from your disk file
   3 --- to type them all in yourself
*** Please type the number of your option >>>????>1

* Standard test/demo requirements to be used
    From file DCB:AC-Requirements-Test.LSP
Requirements:001
```

!!! Note:
There are about 20 values given as requirements,
including the maximum operating temperature and
pressure, and the size of the envelope in which
the air-cylinder must fit. They are as follows.

```
EnvelopeLength      ---- 7.83
EnvelopeHeight      ---- 1.5
EnvelopeWidth       ---- 1.75
MaxTemperature      ---- 250
OperatingMedium     ---- Air
OperatingPressureMax ---- 60
OperatingPressureMin ---- 30
RodLoad             ---- 1.4
Stroke              ---- 1.75
RodThreadType       ---- UNF24
RodThreadLength     ---- 1.031
RodDiameter
       ---- (LNGTH 0.312 0.0 2.e-3)
```

```
Environment        ---- Corrosive
Quality            ---- Reliable
MTBF               ---- 100000
AirInletDiameter   ---- 0.374
MountingScrewSize
        ---- (LNGTH 0.19 5.e-3 5.e-3)
MountingHoleToHole
        ---- (LNGTH 0.625 5.e-3 5.e-3)
MaxFaceToMountingHoles
        ---- (LNGTH 0.31 5.e-3 5.e-3)
```

*** Requirements Input Complete

--- Entering Specialist
   ...AirCylinder...Mode = Design

!!! Note:
The first part of the design plan is to check
the requirements and then do a rough-design.

----- Entering Specialist
   ...AirCylinder...Mode = RoughDesign

------- Entering Specialist
   ...Head...Mode = RoughDesign

------- Leaving Specialist
   ....Head...Result= Success Msg

------- Entering Specialist
   ...Rest...Mode = RoughDesign

--------- Entering Specialist
   ...Cap...Mode = RoughDesign

--------- Leaving Specialist
   ....Cap...Result= Success Msg

--------- Entering Specialist
   ...Tube...Mode = RoughDesign

!!! Note:
At any point the system can ask the user for
a value, or ask for a value to be checked.
This is useful in situations where it is hard to
capture the judgment being made, or in
development situations where that part of
the design has yet has yet to be codified.

???? ASK-USER: Tube rough volume is 4.051
         Is this OK for this design?
Answer is ????>yes

--------- Leaving Specialist
       ....Tube...Result= Success Msg

--------- Entering Specialist
       ...PistonAndRod... Mode = RoughDesign

--------- Leaving Specialist
       ....PistonAndRod...Result= Success Msg

--------- Entering Specialist
       ...Bumper... Mode = RoughDesign

--------- Leaving Specialist
       ....Bumper...Result= Success Msg

------- Leaving Specialist
       ....Rest...Result= Success Msg

------- Entering Specialist
       ...Spring... Mode = RoughDesign

------- Leaving Specialist
       ....Spring...Result= Success Msg

----- Leaving Specialist
       ....AirCylinder...Result= Success Msg

!!! Note:
The rough design is complete.
Now do the design.

----- Entering Specialist
       ...Spring... Mode = Design

???? ASK-VALUE: Spring Wire Diameter
Value is ????>.215

???? ASK-VALUE: Number of Coils
Value is ????>11

!!! Note:
Every plan can include a report
of its results. As the spring specialist
is largely a dummy it produces few results.

```
SpringMaterial         ---- NIL
SpringOD               ---- 0.985
SpringID               ---- 0.77
SpringWireDiameter     ---- 0.215
SpringFreeLength       ---- NIL
SpringCompressedLength ---- NIL
SpringInstalledLength  ---- NIL
SpringLoad             ---- NIL
SpringNumberOfCoils    ---- 11
SpringDeflectionPerCoil ---- NIL
```

----- Leaving Specialist
    ....Spring...Result= Success Msg

----- Entering Specialist
    ...Head... Mode = Design

???? ASK-USER:
Head Air Cavity volume is 0.323
Is this OK for this design?
Answer is ????>ok

!!! Note:
The LNGTH form below is a way of expressing
tolerances. The first figure is the value,
the second the +ve tolerance, while the third
is the -ve tolerance.

```
HeadWidth     ---- 1.5
HeadDepth     ---- 0.97
HeadHeight    ---- 1.5
HeadMaterial  ---- StainlessSteel
HeadScrewSize ---- (LNGTH 0.19 5.e-3 5.e-3)
HeadCenterCenterDistance
              ---- (LNGTH 0.625 5.e-3 5.e-3)
HeadMountingHoleDiameter
              ---- (LNGTH 0.206 3.e-3 0.0)
HeadCounterSinkDiameter
              ---- (LNGTH 0.37 1.e-2 1.e-2)
HeadMaxHtoFDistance
              ---- (LNGTH 0.31 5.e-3 5.e-3)
HeadMountingHolesToFaceDistance
              ---- (LNGTH 0.2455 2.5e-3 2.5e-3)
HeadWiperSeatDepth      ---- 0.175854
HeadWiperSeatDiameter   ---- 0.459841
HeadWiperType           ---- UCup
HeadAirHoleToSideDistance ---- 0.75
HeadAirHoleToFaceDistance ---- 0.701
```

HeadAirHoleDepth ---- 0.2105
HeadAirHoleDiameter ---- 0.374
HeadAirCavityID ---- 0.534
HeadAirCavityOD ---- 1.089
HeadAirCavityDepth ---- 0.4565
HeadTRHCenterCenterDistance ---- 1.115
HeadTRHDepth ---- NIL
HeadTRHDiameter ---- 0.19
HeadBearingThickness ---- 4.85e-2
HeadBearing1Length ---- 0.4765
HeadBearing2Length ---- 0.182646
HeadSealSeatWidth
    ---- (LNGTH 0.125 5.e-3 5.e-3)
HeadSealSeatToFaceDistance ---- 0.3585
HeadSealSeatDiameter
    ---- (LNGTH 0.5 3.e-3 0.0)
HeadTubeSeatID
    ---- (LNGTH 1.21 6.e-3 3.e-3)
HeadTubeSeatOD
    ---- (LNGTH 1.359 1.e-2 1.e-2)
HeadTubeSeatDepth ---- 6.25d-2

!!! Note:
The system contains a table of standard decimal values
and is able to take the nearest higher or lower value,
or just the nearest. For example, a value of 2.4936 can
be stored as 2.5, or as 2.4844 (ie. 31/64ths).

----- Leaving Specialist
    ....Head...Result= Success Msg

!!! Note:
Once the Head specialist is completed the
Rest specialist can start.

----- Entering Specialist
    ...Rest... Mode = Design

------- Entering Specialist
    ...PistonAndRod... Mode = Design

PistonDiameter
    ---- (LNGTH 1.212 4.e-3 0.0)
PistonMaterial ---- Brass
PistonThickness ---- 0.34375d
PistonRodHole ---- 0.25d
PistonSpringSeatDepth ---- 3.9e-2
PistonSpringSeatID ---- 0.754375

```
PistonSpringSeatOD      ---- 1.00062
PistonSealType          ---- UCup
PistonSealSeatDiameter
     ---- (LNGTH 0.885 0.0 1.e-3 ThreeDP)
PistonSealSeatWidth
     ---- (LNGTH 0.156 1.e-2 1.e-2 ThreeDP)
PistonSealSeatPosition  ---- 9.4e-2
PistonBreakawayCutDiameter ---- 0.729
PistonBreakawayCutDepth    ---- 3.4e-2
PistonNotchCount        ---- 4
PistonNotchWidth        ---- 7.8e-2
PistonNotchDepth        ---- 7.8e-2
PistonBrazeSeatDepth    ---- 7.8e-2
PistonBrazeSeatDiameter ---- 0.390625d

RodDiameter ---- (LNGTH 0.312 0.0 2.e-3)
RodLength               ---- 4.095
RodThreadLength         ---- 1.031
RodThreadType           ---- UNF24
RodMaterial             ---- StainlessSteel
RodPistonSeatDiameter   ---- 0.247
RodPistonSeatLength     ---- 0.31
RodEndOfRodToHead       ---- 2.781

------- Leaving Specialist
     ....PistonAndRod...Result= Success Msg

------- Entering Specialist
     ...Cap... Mode = Design

CapMaterial        ---- StainlessSteel
CapHeight          ---- 1.5
CapWidth           ---- 1.5
CapDepth           ---- 0.625
CapInternalDepth   ---- 0.499
CapInternalDiameter ---- 1.089
CapTubeSeatDepth   ---- 6.25d-2
CapTubeSeatID
     ---- (LNGTH 1.21 6.e-3 3.e-3)
CapTubeSeatOD
     ---- (LNGTH 1.359 1.e-2 1.e-2)
CapAirHoleDiameter ---- 0.374
CapAirHoleCenterToBackDistance
          ---- 0.313
CapAirHoleDepth    ---- 0.2055
CapBackFaceThickness ---- NIL
CapTRtoTRDistance  --- 1.115
CapTRDiameter
```

```
            ---- (LNGTH 0.203 5.e-3 5.e-3)
CapTRDepth            ---- 0.3125
CapTRRecessDepth      ---- 0.3125
CapTRRecessRadius     ---- 1.003
CapLargeChamferWidth  ---- NIL
CapLargeChamferAngle  ---- NIL
CapSmallChamferWidth  ---- NIL
CapSmallChamferAngle  ---- NIL


------- Leaving Specialist
    ....Cap...Result= Success Msg


------- Entering Specialist
    ...Tube... Mode = Design

TubeMaterial         ---- StainlessSteel
TubeLength           ---- 3.5
TubeID               ---- 1.214
TubeOD               ---- 1.344
TubeChamferLength    ---- NIL
TubeChamferAngle     ---- NIL


------- Leaving Specialist
    ....Tube...Result= Success Msg


------- Entering Specialist
    ...Bumper... Mode = Design

BumperMaterial        ---- StainlessSteel
BumperLength          ---- NIL
BumperID              ---- 0.390625d
BumperOD              ---- 0.69
BumperFlangeDiameter  ---- 1.059
BumperFlangeThickness ---- 6.25e-2


------- Leaving Specialist
    ....Bumper...Result= Success Msg

----- Leaving Specialist
    ....Rest...Result= Success Msg

--- Leaving Specialist
    ....AirCylinder...Result= Success Msg

*** Design attempt succeeds
*** Version date: (4 26 84)
*** Todays date:  (5 4 84)
*** User: DCBROWN
```

Appendix B: Design Trace with Step Redesign

This is a trace generated by the system. Much of it has been edited out for brevity, and for presentation in this format. The trace was turned on for all agents. The trace is of a successful design with step redesign and selection of alternative plans.

***** AIR-CYL Air-cylinder Design System *****
*** Version date: (4 26 84)
*** Todays date: (5 1 84)
*** User: DCBROWN

*** Requirements input
The following options are available:
   1 --- to use a set of standard test/demo requirements
   2 --- to have requirements read from your disk file
   3 --- to type them all in yourself
Note: you will be able to make alterations
*** Please type the number of your option >>>????>1

* Standard test/demo requirements to be used
     From file DCB:AC-Requirements-Test.LSP
Requirements:001

* Do you wish to make alterations to the requirements?
     Please answer YES or NO or QUIT >>>????>y

EnvelopeLength   ---- 7.83
EnvelopeHeight   ---- 1.5
EnvelopeWidth   ---- 1.75
MaxTemperature   ---- 250
OperatingMedium   ---- Air
OperatingPressureMax ---- 60
OperatingPressureMin ---- 30
RodLoad   ---- 1.4
Stroke   ---- 1.75
RodThreadType   ---- UNF24
RodThreadLength   ---- 1.031
RodDiameter
     ---- (LNGTH 0.312 0.0 2.e-3)
Environment   ---- Corrosive
Quality   ---- Reliable

```
MTBF              ---- 100000
AirInletDiameter  ---- 0.374
MountingScrewSize
      ---- (LNGTH 0.19 5.e-3 5.e-3)
MountingHoleToHole
      ---- (LNGTH 0.625 5.e-3 5.e-3)
MaxFaceToMountingHoles
      ---- (LNGTH 0.31 5.e-3 5.e-3)

* Alterations from user
  Use EXIT or QUIT
     to show that you have finished
  Use HELP
     to get list of system names for requirements
  Use SHOW
     to see current state of the requirements
System name for requirement is >>>????>EnvelopeWidth
 "Envelope width"
Is this the correct requirement? >>>????>y

     Current value is 1.75

     New value is >>>????>1.35

!!! Note:
We have cut down the width of the envelope
without altering any other requirement
in order to make the design harder.

     System name for requirement is >>>????>quit

* End of alterations from user
*** Requirements Input Complete

--- Entering Specialist
    ...AirCylinder... Mode = Design

----- Entering Plan
      ...AirCylinderDP1... Type = Design

------- Entering Task
        ...CheckRequirements

--------- Entering Step
          ...CheckEnvelope

--------- Leaving Step
          ....CheckEnvelope...Result= Success Msg
```

!!! etc !!!

--------- Leaving Plan
   ....AirCylinderRDP1...Result= Success Msg

------- Leaving Specialist
   ....AirCylinder...Result= Success Msg

!!! etc !!!

------- Entering Specialist
   ...Rest... Mode = Design

--------- Entering Plan
   ...RestDP1... Type = Design

----------- Entering Specialist
   ...PistonAndRod... Mode = Design

!!! Note:
At this point the system is working on the
design of the piston and rod assembly.
This is where the trouble starts.

------------- Entering Plan
   ...PistonAndRodDP1... Type = Design

!!! etc !!!

--------------- Entering Task
   ...PistonSeal

---------------- Entering Step
   ...PistonSealType

---------------- Leaving Step
   ....PistonSealType
   ...Result= Success Msg

---------------- Entering Step
   ...PistonSealSeatWidth

!!! Note:
The constraint test that follows will
discover that there isn't enough space
in the piston for the seat for the seal
that will go around the piston. Its
failure produces a message which shows

in detail how the failure occurred.

```
------------------ Entering TEST-CONSTRAINTS
           ...(Available>Width)

------------------ Leaving TEST-CONSTRAINTS
           ....(Available>Width)...Result=
```

#S(Msg Msg:MsgType Failure Msg:MsgSubType Constraint Msg:FromName Available>Width Msg:FromType Constraint Msg:Message "Constraint failure" Msg:Explanation "Seal width is greater than available space in piston" Msg:ToName NIL Msg:ToType NIL Msg:Destination NIL Msg:InPlan PistonAndRodDP1 Msg:InMode Design Msg:Attitude Normal Msg:Constraints NIL Msg:SuggestionsBelow NIL Msg:Suggestions (#S(Suggestion Suggestion:FromName Available>Width Suggestion:FromType Constraint Suggestion:AttributeName PistonThickness Suggestion:SuccessIf (Width <= 0.140875011682510376d) Suggestion:FailingValue (LNGTH 0.156 1.e-2 1.e-2 ThreeDP) Suggestion:InPlan PistonAndRodDP1 Suggestion:Suggest (INCREASE PistonThickness BY 1.5124976634979248d-2)) #S(Suggestion Suggestion:FromName Available>Width Suggestion:FromType Constraint Suggestion:AttributeName PistonSealSeatWidth Suggestion:SuccessIf (Width <= 0.140875011682510376d) Suggestion:FailingValue (LNGTH 0.156 1.e-2 1.e-2 ThreeDP) Suggestion:InPlan PistonAndRodDP1 Suggestion:Suggest (DECREASE PistonSealSeatWidth BY 1.5124976634979248d-2))) Msg:ContributingMsg #S(Msg Msg:MsgType Failure Msg:MsgSubType Agent Msg:FromName Available>Width Msg:FromType Constraint Msg:Message "Constraint TEST is False" Msg:Explanation (TEST (Width <= Available)) Msg:ToName NIL Msg:ToType NIL Msg:Destination NIL Msg:InPlan PistonAndRodDP1 Msg:InMode Design Msg:Attitude Normal Msg:Constraints NIL Msg:SuggestionsBelow NIL Msg:Suggestions NIL Msg:ContributingMsg NIL))

!!! Note:
As a result of the constraint failing the
Decision part of the step (containing the
constraint test) will fail too.

```
------------------ Leaving DECISION
           ... Result=
```

#S(Msg Msg:MsgType Failure Msg:MsgSubType Agent Msg:FromName PistonSealSeatWidth Msg:FromType Step Msg:Message "Decision -- constraint failure" Msg:Explanation NIL Msg:ToName NIL Msg:ToType NIL Msg:Destination NIL Msg:InPlan PistonAndRodDP1 Msg:InMode Design Msg:Attitude Normal Msg:Constraints NIL .....

!!! etc !!!

```
------------------ Entering FailureHandler
           ...SystemRealStepFH

-------------------- Entering FailureHandler
```

```
                ...SystemStepBodyFailureFH

---------------------- Entering FailureHandler
                ...SystemDecisionFH

------------------------ Entering FailureHandler
                ...SystemDecisionConstraintFH
```

!!! Note:
The failure handlers for a step which are
built into the system determine that a domain
specific (i.e., USER) failure handler will be
able to decide what to do. Notice that the
failure handlers home in on the problem
by looking at the nested failure messages.

```
-------------------------- Entering FailureHandler
                ...PistonSealSeatWidthFH
```

!!! Note:
The failure handler says to try redesign.

```
---------------------------- Entering Redesigner
                ...PistonSSWRedesigner
        Step = PistonSealSeatWidth
        Suggest = (DECREASE PistonSealSeatWidth
                BY 1.51749766373541206d-2)

---------------------------- Leaving Redesigner
                ....PistonSSWRedesigner
                ...Result= Success Msg
```

!!! Note:
The piston seal seat width redesigner
was able to decrease the width as
suggested. Now we go back up the decision
tree of failure handlers with the result.

```
-------------------------- Leaving FailureHandler
                ....PistonSealSeatWidthFH
                ...Result= Success Msg

------------------------ Leaving FailureHandler
                ....SystemDecisionConstraintFH
                ...Result= Success Msg

---------------------- Leaving FailureHandler
```

```
              ....SystemDecisionFH
                ...Result= Success Msg
```

```
-------------------- Leaving FailureHandler
              ....SystemStepBodyFailureFH
                ...Result= Success Msg
```

```
------------------ Leaving FailureHandler
              ....SystemRealStepFH
                ...Result= Success Msg
```

!!! Note:
We leave the failure handlers and
return to the step.
The redesign was successful,
so the step is successful and
acts as if no problems were
encountered.

```
----------------- Leaving Step
              ....PistonSealSeatWidth
                ...Result= Success Msg
```

   !!! etc !!!

```
------------- Leaving Plan
           ....PistonAndRodDP1
            ...Result= Success Msg
```

```
----------- Leaving Specialist
          ....PistonAndRod...Result= Success Msg
```

```
----------- Entering Specialist
           ...Cap... Mode = Design
```

!!! Note:
Now we attempt design of the cap,
and discover another problem.

```
------------- Entering Plan
          ...CapDP1... Type = Design
```

   !!! etc !!!

```
--------------- Entering Task
             ...CapInternal
```

```
---------------- Entering Step
```

...CapInternalDiameter

!!! Note:
The constraint test to see if the internal diameter of the cap is larger than the outside diameter of the spring, as one must fit in the other. It fails.

------------------ Entering TEST-CONSTRAINTS
...(CapID>SpringOD)

------------------ Leaving TEST-CONSTRAINTS
....(CapID>SpringOD)...Result=

#S(Msg Msg:MsgType Failure Msg:MsgSubType Constraint Msg:FromName CapID>SpringOD Msg:FromType Constraint Msg:Message "Constraint failure" Msg:Explanation "Cap internal diameter too small for spring" Msg:ToName NIL Msg:ToType NIL Msg:Destination NIL Msg:InPlan CapDP1 Msg:InMode Design Msg:Attitude Normal Msg:Constraints NIL Msg:SuggestionsBelow NIL Msg:Suggestions (#S(Suggestion Suggestion:FromName CapID>SpringOD Suggestion:FromType Constraint Suggestion:AttributeName SpringOD Suggestion:SuccessIf (InternalDiameter > 0.985) Suggestion:FailingValue 0.886 Suggestion:InPlan CapDP1 Suggestion:Suggest (DECREASE SpringOD BY 9.90001e-2)) #S(Suggestion Suggestion:FromName CapID>SpringOD Suggestion:FromType Constraint Suggestion:AttributeName CapInternalDiameter Suggestion:SuccessIf (InternalDiameter > 0.985) Suggestion:FailingValue 0.886 Suggestion:InPlan CapDP1 Suggestion:Suggest (INCREASE CapInternalDiameter BY 9.90001e-2))) Msg:ContributingMsg #S(Msg Msg:MsgType Failure Msg:MsgSubType Agent Msg:FromName CapID>SpringOD Msg:FromType Constraint Msg:Message "Constraint TEST is False" Msg:Explanation (TEST (InternalDiameter > SpringOD)) Msg:ToName NIL Msg:ToType NIL Msg:Destination NIL Msg:InPlan CapDP1 Msg:InMode Design Msg:Attitude Normal Msg:Constraints NIL Msg:SuggestionsBelow NIL Msg:Suggestions NIL Msg:ContributingMsg NIL))

------------------ Leaving DECISION
...Result=

#S(Msg Msg:MsgType Failure Msg:MsgSubType Agent Msg:FromName CapInternalDiameter Msg:FromType Step Msg:Message "Decision -- constraint failure" Msg:Explanation NIL Msg:ToName NIL Msg:ToType NIL Msg:Destination NIL Msg:InPlan CapDP1 ......

!!! etc !!!

------------------ Entering FailureHandler
...SystemRealStepFH

-------------------- Entering FailureHandler
...SystemStepBodyFailureFH

```
----------------------- Entering FailureHandler
                ...SystemDecisionFH

----------------------- Entering FailureHandler
                ...SystemDecisionConstraintFH

------------------------- Entering FailureHandler
                ...CapIDFH
```

!!! Note:
The domain specific failure handler
says to try redesign.

```
--------------------------- Entering Redesigner
                ...CapIDRedesigner
        Step = CapInternalDiameter
        Suggest = (INCREASE CapInternalDiameter
                BY 9.90501e-2)

----------------------------- Entering TEST-CONSTRAINT
                ...(CapID>SpringOD)

----------------------------- Leaving TEST-CONSTRAINT
                ....(CapID>SpringOD)
                ...Result= Success Msg

--------------------------- Leaving Redesigner
                ....CapIDRedesigner
                ...Result= Success Msg
```

!!! Note:
The redesign is successful.
The suggested increase could be made,
and the constraint was satisfied.

```
--------------------------- Leaving FailureHandler
                ....CapIDFH
                ...Result= Success Msg

------------------------ Leaving FailureHandler
                ....SystemDecisionConstraintFH
                ...Result= Success Msg

---------------------- Leaving FailureHandler
                ....SystemDecisionFH
                ...Result= Success Msg

-------------------- Leaving FailureHandler
```

                    ....SystemStepBodyFailureFH
                    ...Result= Success Msg

------------------ Leaving FailureHandler
                    ....SystemRealStepFH
                    ...Result= Success Msg

!!! Note:
The step is successful,
as the failure was handled.

----------------- Leaving Step
                    ....CapInternalDiameter
                    ...Result= Success Msg

   !!! etc !!!

------------- Leaving Plan
            ....CapDP1...Result= Success Msg

----------- Leaving Specialist
            ....Cap...Result= Success Msg

   !!! etc !!!

----------- Entering Specialist
            ...Bumper... Mode = Design

!!! Note:
The bumper is being designed here.
More problems are encountered.

------------- Entering Plan
            ...BumperDP1... Type = Design

--------------- Entering Task
            ...BumperFlange

----------------- Entering Step
            ...BumperFlangeDiameter

!!! Note:
The bumper flange diameter must
be large enough to support the spring.
The constraint tests that, but fails.

------------------ Entering TEST-CONSTRAINTS
            ...(BFD>SpringOD)

------------------- Leaving TEST-CONSTRAINTS
        ....(BFD>SpringOD)
       ...Result=

#S(Msg Msg:MsgType Failure Msg:MsgSubType Constraint Msg:FromName BFD>SpringOD Msg:FromType Constraint Msg:Message "Constraint failure" Msg:Explanation "Bumper flange is too small for spring" Msg:ToName NIL Msg:ToType NIL Msg:Destination NIL Msg:InPlan BumperDP1 Msg:InMode Design Msg:Attitude Normal Msg:Constraints NIL Msg:SuggestionsBelow NIL Msg:Suggestions (#S(Suggestion Suggestion:FromName BFD>SpringOD Suggestion:FromType Constraint Suggestion:AttributeName SpringOD Suggestion:SuccessIf (FlangeDiameter >= 0.985) Suggestion:FailingValue 0.95505 Suggestion:InPlan BumperDP1 Suggestion:Suggest (DECREASE SpringOD BY 2.99501e-2)) #S(Suggestion Suggestion:FromName BFD>SpringOD Suggestion:FromType Constraint Suggestion:AttributeName BumperFlangeDiameter Suggestion:SuccessIf (FlangeDiameter >= 0.985) Suggestion:FailingValue 0.95505 Suggestion:InPlan BumperDP1 Suggestion:Suggest (INCREASE BumperFlangeDiameter BY 2.99501e-2))) Msg:ContributingMsg #S(Msg Msg:MsgType Failure Msg:MsgSubType Agent Msg:FromName BFD>SpringOD Msg:FromType Constraint Msg:Message "Constraint TEST is False" Msg:Explanation (TEST (FlangeDiameter >= SpringOD)) Msg:ToName NIL Msg:ToType NIL Msg:Destination NIL Msg:InPlan BumperDP1 Msg:InMode Design Msg:Attitude Normal Msg:Constraints NIL Msg:SuggestionsBelow NIL Msg:Suggestions NIL Msg:ContributingMsg NIL))

!!! Note:
The decision section of the step
fails as a consequence.

------------------- Leaving DECISION
       ...Result=

#S(Msg Msg:MsgType Failure Msg:MsgSubType Agent Msg:FromName BumperFlangeDiameter Msg:FromType Step Msg:Message "Decision -- constraint failure" Msg:Explanation NIL Msg:ToName NIL Msg:ToType NIL Msg:Destination NIL
....

    !!! etc !!!

------------------- Entering FailureHandler
       ...SystemRealStepFH

--------------------- Entering FailureHandler
       ...SystemStepBodyFailureFH

---------------------- Entering FailureHandler
       ...SystemDecisionFH

------------------------ Entering FailureHandler
       ...SystemDecisionConstraintFH

-------------------------- Entering FailureHandler
                 ...BumperFDFH

!!! Note:
The domain specific failure handler
says to try redesign.

--------------------------- Entering Redesigner
                 ...BumperFDRedesigner
     Step = BumperFlangeDiameter
     Suggest = (INCREASE BumperFlangeDiameter
          BY 3.00001e-2)

----------------------------- Leaving DECISION
                 ...Result=

#S(Msg Msg:MsgType Failure Msg:MsgSubType Agent Msg:FromName BumperFDRedesigner Msg:FromType Redesigner Msg:Message "Increase not possible in redesigner" Msg:Explanation (INCREASE BumperFlangeDiameter BY 3.00001e-2) Msg:ToName NIL Msg:ToType NIL Msg:Destination NIL Msg:InPlan BumperDP1 Msg:InMode Design Msg:Attitude Normal Msg:Constraints NIL Msg:SuggestionsBelow NIL Msg:Suggestions NIL Msg:ContributingMsg NIL)

!!! Note:
The decision part of the redesigner fails
as there is no knowledge there about
increasing the value of that attribute.
The redesigner fails.

--------------------------- Leaving Redesigner
                 ....BumperFDRedesigner
                 ...Result=

#S(Msg Msg:MsgType Failure Msg:MsgSubType Agent Msg:FromName BumperFDRedesigner Msg:FromType Redesigner Msg:Message "Redesigner action section fails" Msg:Explanation (INCREASE BumperFlangeDiameter BY 3.00001e-2) Msg:ToName NIL Msg:ToType NIL Msg:Destination NIL Msg:InPlan BumperDP1 Msg:InMode Design Msg:Attitude Normal Msg:Constraints NIL Msg:SuggestionsBelow NIL Msg:Suggestions NIL Msg:ContributingMsg #S(Msg Msg:MsgType Failure Msg:MsgSubType Agent Msg:FromName BumperFDRedesigner Msg:FromType Redesigner Msg:Message "Decision failure" Msg:Explanation NIL Msg:ToName NIL Msg:ToType NIL Msg:Destination NIL Msg:InPlan BumperDP1 Msg:InMode Design Msg:Attitude Normal Msg:Constraints NIL Msg:SuggestionsBelow NIL Msg:Suggestions NIL Msg:ContributingMsg #S(Msg Msg:MsgType Failure Msg:MsgSubType Agent Msg:FromName BumperFDRedesigner Msg:FromType Redesigner Msg:Message "Increase not possible in redesigner" Msg:Explanation (INCREASE BumperFlangeDiameter BY 3.00001e-2) Msg:ToName NIL Msg:ToType NIL Msg:Destination NIL Msg:InPlan BumperDP1 Msg:InMode Design Msg:Attitude Normal Msg:Constraints NIL Msg:SuggestionsBelow NIL Msg:Suggestions NIL

Msg:ContributingMsg NIL)))

!!! Note:
All the failure handlers report failure
to the failure handler above, and
eventually the step gets told of the bad
news.

-------------------------- Leaving FailureHandler
          ....BumperFDFH...Result=

#S(Msg Msg:MsgType Failure Msg:MsgSubType Agent Msg:FromName BumperFDRedesigner Msg:FromType Redesigner Msg:Message "Redesigner action section fails" Msg:Explanation (INCREASE BumperFlangeDiameter BY 3.00001e-2) Msg:ToName NIL Msg:ToType NIL Msg:Destination NIL Msg:InPlan BumperDP1 Msg:InMode Design Msg:Attitude ......

   !!! etc !!!

------------------- Leaving FailureHandler
          ....SystemRealStepFH...Result=

#S(Msg Msg:MsgType Failure Msg:MsgSubType Agent Msg:FromName BumperFDRedesigner Msg:FromType Redesigner Msg:Message "Redesigner action section fails" Msg:Explanation (INCREASE BumperFlangeDiameter BY 3.00001e-2) Msg:ToName NIL Msg:ToType NIL Msg:Destination NIL Msg:InPlan BumperDP1 Msg:InMode Design Msg:Attitude Normal Msg:Constraints NIL Msg:SuggestionsBelow NIL Msg:Suggestions NIL Msg:ContributingMsg .....

!!! Note:
The whole failure message chain for
the step is shown below. Embedded messages
begin with a "#S(Msg".

----------------- Leaving Step
          ....BumperFlangeDiameter
        ...Result=

#S(Msg Msg:MsgType Failure Msg:MsgSubType Agent Msg:FromName BumperFlangeDiameter Msg:FromType Step Msg:Message "Step failure" Msg:Explanation #S(Msg Msg:MsgType Failure Msg:MsgSubType Agent Msg:FromName BumperFDRedesigner Msg:FromType Redesigner Msg:Message "Redesigner action section fails" Msg:Explanation (INCREASE BumperFlangeDiameter BY 3.00001e-2) Msg:ToName NIL Msg:ToType NIL Msg:Destination NIL Msg:InPlan BumperDP1 Msg:InMode Design Msg:Attitude Normal Msg:Constraints NIL Msg:SuggestionsBelow NIL Msg:Suggestions NIL Msg:ContributingMsg #S(Msg Msg:MsgType Failure Msg:MsgSubType Agent Msg:FromName BumperFDRedesigner Msg:FromType Redesigner Msg:Message "Decision failure" Msg:Explanation NIL Msg:ToName NIL Msg:ToType NIL Msg:Destination NIL Msg:InPlan BumperDP1 Msg:InMode Design Msg:Attitude

Normal Msg:Constraints NIL Msg:SuggestionsBelow NIL Msg:Suggestions NIL Msg:ContributingMsg #S(Msg Msg:MsgType Failure Msg:MsgSubType Agent Msg:FromName BumperFDRedesigner Msg:FromType Redesigner Msg:Message "Increase not possible in redesigner" Msg:Explanation (INCREASE BumperFlangeDiameter BY 3.00001e-2) Msg:ToName NIL Msg:ToType NIL Msg:Destination NIL Msg:InPlan BumperDP1 Msg:InMode Design Msg:Attitude Normal Msg:Constraints NIL Msg:SuggestionsBelow NIL Msg:Suggestions NIL Msg:ContributingMsg NIL))) Msg:ToName NIL Msg:ToType NIL Msg:Destination NIL Msg:InPlan BumperDP1 Msg:InMode Design Msg:Attitude Normal Msg:Constraints NIL Msg:SuggestionsBelow "Suggestions below" Msg:Suggestions (#S(Suggestion Suggestion:FromName BumperFlangeDiameter Suggestion:FromType Step Suggestion:AttributeName CapInternalDiameter Suggestion:SuccessIf NIL Suggestion:FailingValue NIL Suggestion:InPlan BumperDP1 Suggestion:Suggest (INCREASE CapInternalDiameter)) #S(Suggestion Suggestion:FromName BumperFlangeDiameter Suggestion:FromType Step Suggestion:AttributeName SpringOD Suggestion:SuccessIf NIL Suggestion:FailingValue NIL Suggestion:InPlan BumperDP1 Suggestion:Suggest (DECREASE SpringOD))) Msg:ContributingMsg

#S(Msg Msg:MsgType Failure Msg:MsgSubType Agent Msg:FromName BumperFlangeDiameter Msg:FromType Step Msg:Message "Step body failure" Msg:Explanation NIL Msg:ToName NIL Msg:ToType NIL Msg:Destination NIL Msg:InPlan BumperDP1 Msg:InMode Design Msg:Attitude Normal Msg:Constraints NIL Msg:SuggestionsBelow "Suggestions below" Msg:Suggestions NIL Msg:ContributingMsg

#S(Msg Msg:MsgType Failure Msg:MsgSubType Agent Msg:FromName BumperFlangeDiameter Msg:FromType Step Msg:Message "Decision failure" Msg:Explanation NIL Msg:ToName NIL Msg:ToType NIL Msg:Destination NIL Msg:InPlan BumperDP1 Msg:InMode Design Msg:Attitude Normal Msg:Constraints NIL Msg:SuggestionsBelow "Suggestions below" Msg:Suggestions NIL Msg:ContributingMsg

#S(Msg Msg:MsgType Failure Msg:MsgSubType Agent Msg:FromName BumperFlangeDiameter Msg:FromType Step Msg:Message "Decision -- constraint failure" Msg:Explanation NIL Msg:ToName NIL Msg:ToType NIL Msg:Destination NIL Msg:InPlan BumperDP1 Msg:InMode Design Msg:Attitude Normal Msg:Constraints NIL Msg:SuggestionsBelow "Suggestions below" Msg:Suggestions NIL Msg:ContributingMsg

#S(Msg Msg:MsgType Failure Msg:MsgSubType Constraint Msg:FromName BFD>SpringOD Msg:FromType Constraint Msg:Message "Constraint failure" Msg:Explanation "Bumper flange is too small for spring" Msg:ToName NIL Msg:ToType NIL Msg:Destination NIL Msg:InPlan BumperDP1 Msg:InMode Design Msg:Attitude Normal Msg:Constraints NIL Msg:SuggestionsBelow NIL Msg:Suggestions (#S(Suggestion Suggestion:FromName BFD>SpringOD Suggestion:FromType Constraint Suggestion:AttributeName SpringOD Suggestion:SuccessIf (FlangeDiameter >= 0.985) Suggestion:FailingValue 0.95505 Suggestion:InPlan BumperDP1 Suggestion:Suggest (DECREASE SpringOD BY 2.99501e-2)) #S(Suggestion Suggestion:FromName BFD>SpringOD Suggestion:FromType Constraint Suggestion:AttributeName BumperFlangeDiameter Suggestion:SuccessIf (FlangeDiameter >= 0.985) Suggestion:FailingValue 0.95505 Suggestion:InPlan BumperDP1 Suggestion:Suggest (INCREASE

BumperFlangeDiameter BY 2.99501e-2))) Msg:ContributingMsg
#S(Msg Msg:MsgType Failure Msg:MsgSubType Agent Msg:FromName BFD>SpringOD Msg:FromType Constraint Msg:Message "Constraint TEST is False" Msg:Explanation (TEST (FlangeDiameter >= SpringOD)) Msg:ToName NIL Msg:ToType NIL Msg:Destination NIL Msg:InPlan BumperDP1 Msg:InMode Design Msg:Attitude Normal Msg:Constraints NIL Msg:SuggestionsBelow NIL Msg:Suggestions NIL Msg:ContributingMsg NIL))))))

!!! Note:
The step has failed. The task
passes the failure message from
the step to its failure handler.
Other failure handlers will attempt
to determine if the task can do
anything about the failure reported.

----------------- Entering FailureHandler
　　　　...SystemTaskBodyFailureFH

------------------ Entering FailureHandler
　　　　...BumperFlangeFH

!!! Note:
The failure handler for the task
discovers that no suggestions
have been passed up from below.
This means that no redesign can be
considered. The failure handlers
fail as they couldn't handle the
problem.

------------------ Leaving FailureHandler
　　　　....BumperFlangeFH
　　　　...Result=

#S(Msg Msg:MsgType Failure Msg:MsgSubType Agent Msg:FromName BumperFlangeFH Msg:FromType FailureHandler Msg:Message "No relevant suggestions for task redesigner" Msg:Explanation NIL Msg:ToName NIL Msg:ToType NIL Msg:Destination NIL Msg:InPlan BumperDP1 Msg:InMode Design Msg:Attitude Normal Msg:Constraints NIL Msg:SuggestionsBelow NIL Msg:Suggestions NIL Msg:ContributingMsg NIL)

----------------- Leaving FailureHandler
　　　　....SystemTaskBodyFailureFH
　　　　...Result=

#S(Msg Msg:MsgType Failure Msg:MsgSubType Agent Msg:FromName BumperFlangeFH Msg:FromType FailureHandler Msg:Message "No relevant suggestions for task redesigner" Msg:Explanation NIL Msg:ToName NIL Msg:ToType NIL Msg:Destination NIL Msg:InPlan BumperDP1 Msg:InMode Design Msg:Attitude

Normal Msg:Constraints NIL Msg:SuggestionsBelow NIL Msg:Suggestions NIL Msg:ContributingMsg NIL)

!!! Note:
Upper FH for task is responsible for
guarding the exit from the task.
It catches any failures. In this
case it fails as "Step failure"
has already been the subject of
a recovery attempt.

```
---------------- Entering FailureHandler
          ...SystemRealTaskFH

------------------ Entering FailureHandler
          ...SystemTBFailureFH

------------------ Leaving FailureHandler
          ....SystemTBFailureFH
          ...Result=
```

#S(Msg Msg:MsgType Failure Msg:MsgSubType FailureHandler Msg:FromName SystemTBFailureFH Msg:FromType FailureHandler Msg:Message "FAIL used in failure handler" Msg:Explanation "Step failure" Msg:ToName NIL Msg:ToType NIL Msg:Destination NIL Msg:InPlan BumperDP1 Msg:InMode Design Msg:Attitude Normal Msg:Constraints NIL Msg:SuggestionsBelow NIL Msg:Suggestions NIL Msg:ContributingMsg NIL)

```
---------------- Leaving FailureHandler
          ....SystemRealTaskFH
          ...Result=
```

#S(Msg Msg:MsgType Failure Msg:MsgSubType FailureHandler Msg:FromName SystemTBFailureFH Msg:FromType FailureHandler Msg:Message "FAIL used in failure handler" Msg:Explanation "Step failure" Msg:ToName NIL Msg:ToType NIL Msg:Destination NIL Msg:InPlan BumperDP1 Msg:InMode Design Msg:Attitude Normal Msg:Constraints NIL Msg:SuggestionsBelow NIL Msg:Suggestions NIL Msg:ContributingMsg NIL)

!!! Note:
The step failure and subsequent
failing redesign attempt led to
a failure in the task as there
were no suggestions with which
to try a redesign. The task
fails.

```
--------------- Leaving Task
          ....BumperFlange
          ...Result=
```

#S(Msg Msg:MsgType Failure Msg:MsgSubType Agent Msg:FromName BumperFlange Msg:FromType Task Msg:Message "Task failure" Msg:Explanation #S(Msg Msg:MsgType Failure Msg:MsgSubType FailureHandler Msg:FromName SystemTBFailureFH Msg:FromType FailureHandler Msg:Message "FAIL used in failure handler" Msg:Explanation "Step failure" ......

!!! etc !!!

!!! Note:
And the plan fails due to
the failing task.

-------------- Leaving Plan
            ....BumperDP1
            ...Result=

#S(Msg Msg:MsgType Failure Msg:MsgSubType Agent Msg:FromName BumperDP1 Msg:FromType Plan Msg:Message "Plan failure" Msg:Explanation NIL Msg:ToName NIL Msg:ToType NIL Msg:Destination NIL Msg:InPlan BumperDP1 Msg:InMode Design Msg:Attitude Normal Msg:Constraints NIL Msg:SuggestionsBelow NIL Msg:Suggestions NIL Msg:ContributingMsg #S(Msg Msg:MsgType Failure Msg:MsgSubType Agent Msg:FromName BumperFlange Msg:FromType Task Msg:Message "Task failure" ....

!!! etc !!!

!!! Note:
The next plan is selected,
as the last one failed.

-------------- Entering Plan
            ...BumperDP2... Type = Design

--------------- Entering Task
            ...BumperFlange2

----------------- Entering Step
            ...BumperFlangeDiameter2

------------------ Entering TEST-CONSTRAINTS
            ...(BFD<CapID)

!!! Note:
This is the same constraint that
failed in the last plan. This
time it is ok. The step succeeds.

------------------ Leaving TEST-CONSTRAINTS
            ....(BFD<CapID)

```
                    ...Result= Success Msg

----------------- Leaving Step
            ....BumperFlangeDiameter2
            ...Result= Success Msg

    !!! etc !!!

------------- Leaving Plan
        ....BumperDP2
        ...Result= Success Msg

----------- Leaving Specialist
        ....Bumper...Result= Success Msg

--------- Leaving Plan
        ....RestDP1...Result= Success Msg

------- Leaving Specialist
        ....Rest...Result= Success Msg

----- Leaving Plan
      ....AirCylinderDP1...Result= Success Msg

--- Leaving Specialist
    ....AirCylinder...Result= Success Msg

*** Design attempt succeeds
*** Version date: (4 26 84)
*** Todays date:  (5 1 84)
*** User: DCBROWN
***** AIR-CYL Air-cylinder Design System *****
```

Appendix C: Design Trace with Task Redesign

This is a trace generated by the system. Much of it has been edited out for brevity, and presentation in this format. The trace was turned on for all design agents. This trace shows an example of task redesign.

***** AIR-CYL Air-cylinder Design System *****
*** Version date: (4 26 84)
*** Todays date: (5 1 84)
*** User: DCBROWN

*** Requirements input
The following options are available:
  1 --- to use a set of standard test/demo requirements
  2 --- to have requirements read from your disk file
  3 --- to type them all in yourself
Note: you will be able to make alterations
*** Please type the number of your option >>>????>1

* Standard test/demo requirements to be used
    From file DCB:AC-Requirements-Test.LSP
Requirements:001

* Do you wish to make alterations to the requirements?
    Please answer YES or NO or QUIT >>>????>n

* No alterations to be made
*** Requirements Input Complete

--- Entering Specialist
    ...AirCylinder... Mode = Design

----- Entering Plan
    ...AirCylinderDP1... Type = Design

    !!! etc !!!

------- Entering Specialist
    ...Head... Mode = Design

--------- Entering Plan
    ...HeadDP1... Type = Design

!!! etc !!!

----------- Entering Task
        ...AirCavity

!!! Note:
This example centers around the
air cavity in the head. The steps
all succeed.

------------- Entering Step
        ...AirCavityDepth

------------- Leaving Step
        ....AirCavityDepth
        ...Result= Success Msg

------------- Entering Step
        ...AirCavityID

--------------- Entering TEST-CONSTRAINTS
        ...(ACID)

--------------- Leaving TEST-CONSTRAINTS
        ....(ACID)...Result= Success Msg

------------- Leaving Step
        ....AirCavityID...Result= Success Msg

------------- Entering Step
        ...AirCavityOD

--------------- Entering TEST-CONSTRAINTS
        ...(ACOD>ACID)

--------------- Leaving TEST-CONSTRAINTS
        ....(ACOD>ACID)...Result= Success Msg

------------- Leaving Step
        ....AirCavityOD...Result= Success Msg

------------- Entering Step...CheckAirCavity

!!! Note:
Let us suppose that the volume of the
cavity in this situation is extremely
critical. The user will show that the
design is not adequate. The assumption

is that if it isn't adequate then it is
too small.

???? ASK-USER: Head Air Cavity volume is 0.323
         Is this OK for this design?
Answer is ????>no

--------------- Leaving DECISION
         ....Result=

    #S(Msg Msg:MsgType Failure Msg:MsgSubType Agent Msg:FromName CheckAirCavity Msg:FromType Step Msg:Message "ASK-USER" Msg:Explanation no Msg:ToName NIL Msg:ToType NIL Msg:Destination NIL Msg:InPlan HeadDP1 Msg:InMode Design Msg:Attitude Normal Msg:Constraints NIL Msg:SuggestionsBelow NIL Msg:Suggestions NIL Msg:ContributingMsg NIL)

!!! Note:
As a result of the negative answer
the decision part of the step fails.
The step calls its failure handler.

--------------- Entering FailureHandler
         ...SystemRealStepFH

----------------- Entering FailureHandler
         ...SystemStepBodyFailureFH

------------------ Entering FailureHandler
         ...SystemDecisionFH

-------------------- Entering FailureHandler
         ...CheckAirCavityFH

!!! Note:
As this step is merely present to
check the cavity, which it does by
asking the user, redesign is not
possible for the step. All the
failure handler can do is fail.

-------------------- Leaving FailureHandler
         ....CheckAirCavityFH
         ...Result=

#S(Msg Msg:MsgType Failure Msg:MsgSubType FailureHandler Msg:FromName CheckAirCavityFH Msg:FromType FailureHandler Msg:Message "FAIL used in failure handler" Msg:Explanation "ASK-USER" Msg:ToName NIL Msg:ToType NIL Msg:Destination NIL Msg:InPlan HeadDP1 Msg:InMode Design Msg:Attitude Normal Msg:Constraints NIL Msg:SuggestionsBelow NIL Msg:Suggestions NIL

Msg:ContributingMsg NIL)

!!! etc !!!

-------------- Leaving FailureHandler
    ....SystemRealStepFH
    ...Result=

#S(Msg Msg:MsgType Failure Msg:MsgSubType FailureHandler Msg:FromName CheckAirCavityFH Msg:FromType FailureHandler Msg:Message "FAIL used in failure handler" Msg:Explanation "ASK-USER" Msg:ToName NIL Msg:ToType NIL Msg:Destination NIL Msg:InPlan HeadDP1 Msg:InMode Design Msg:Attitude Normal Msg:Constraints NIL Msg:SuggestionsBelow NIL Msg:Suggestions NIL Msg:ContributingMsg NIL)

!!! Note:
The step will now fail.

------------- Leaving Step
    ....CheckAirCavity
    ...Result=

#S(Msg Msg:MsgType Failure Msg:MsgSubType Agent Msg:FromName CheckAirCavity Msg:FromType Step Msg:Message "Step failure" Msg:Explanation #S(Msg Msg:MsgType Failure Msg:MsgSubType FailureHandler Msg:FromName CheckAirCavityFH Msg:FromType FailureHandler Msg:Message "FAIL used in failure handler" Msg:Explanation "ASK-USER" ....

!!! Note:
The step fails as redesign wasn't possible.
It makes some suggestions. The task sees
the steps failure and passes the message to
a failure handler.

------------- Entering FailureHandler
    ...SystemTaskBodyFailureFH

-------------- Entering FailureHandler
    ...DefaultUserStepFailureFH

!!! Note:
A default failure handler does the work
as the user didn't specify one. It says
that the task should try redesign.

---------------- Entering Redesigner
    ...DefaultTaskRedesigner...
  Task = AirCavity
  Suggestions = (INCREASE HeadAirCavityDepth)
    (INCREASE HeadAirCavityOD)

                (DECREASE HeadAirCavityID)

!!! Note:
There are three attributes mentioned
in suggestions. The least-backup step
with a suggestion is picked for redesign.
A redesign request is made to the step.

------------------- Entering Step
         ...AirCavityOD...
     Redesign request from task : AirCavity

!!! Note:
The step passes the request to
its redesigner along with the
suggestion.

-------------------- Entering Redesigner
            ...AirCavityODRedesigner
    Step = AirCavityOD
    Suggest = (INCREASE HeadAirCavityOD BY 1.e-3)

---------------------- Entering TEST-CONSTRAINTS
            ...(ACOD>ACID)

---------------------- Leaving TEST-CONSTRAINTS
            ....(ACOD>ACID)
            ...Result= Success Msg

!!! Note:
Despite the success of the constraint
on an intermediate value, the attempt
to redesign by increasing the value
cannot be done. The decision section
of the redesigner, and subsequently the
redesigner itself both fail.

---------------------- Leaving DECISION
            ...Result=

#S(Msg   Msg:MsgType   Failure   Msg:MsgSubType   Agent   Msg:FromName
AirCavityODRedesigner Msg:FromType Redesigner Msg:Message "Increase not
possible in redesigner" Msg:Explanation (INCREASE HeadAirCavityOD BY 1.e-3)
Msg:ToName NIL Msg:ToType NIL Msg:Destination NIL Msg:InPlan HeadDP1
Msg:InMode   Redesign   Msg:Attitude   Normal   Msg:Constraints   NIL
Msg:SuggestionsBelow NIL Msg:Suggestions NIL Msg:ContributingMsg NIL)

-------------------- Leaving Redesigner
            ....AirCavityODRedesigner

```
              ...Result=
```

#S(Msg Msg:MsgType Failure Msg:MsgSubType Agent Msg:FromName AirCavityODRedesigner Msg:FromType Redesigner Msg:Message "Redesigner action section fails" Msg:Explanation (INCREASE HeadAirCavityOD BY 1.e-3) Msg:ToName NIL Msg:ToType NIL Msg:Destination NIL Msg:InPlan HeadDP1 Msg:InMode Redesign ....

!!! Note:
The redesign request from the task
to the step fails.

```
------------------ Leaving Step
            ....AirCavityOD
              ...Result=
```

#S(Msg Msg:MsgType Failure Msg:MsgSubType Agent Msg:FromName AirCavityODRedesigner Msg:FromType Redesigner Msg:Message "Redesigner action section fails" Msg:Explanation (INCREASE HeadAirCavityOD BY 1.e-3) Msg:ToName NIL Msg:ToType NIL Msg:Destination NIL Msg:InPlan HeadDP1 Msg:InMode Redesign ....

!!! Note:
The strategy picks the next best
suggestion as the best suggestion
failed. This will be a slightly
larger backing-up than the last.

```
------------------ Entering Step
            ...AirCavityID
      Redesign request from task :  AirCavity

-------------------- Entering Redesigner
              ...AirCavityIDRedesigner
       Step = AirCavityID
       Suggest = (DECREASE HeadAirCavityID
              BY 1.e-3)

---------------------- Entering TEST-CONSTRAINTS
              ...(ACID)

---------------------- Leaving TEST-CONSTRAINTS
              ....(ACID)
              ...Result= Success Msg
```

!!! Note:
The redesign request to this step
fails as well. It just can't
make the requested decrease.
The redesign request of the step

will fail.

```
---------------------- Leaving DECISION
            ...Result=
```

#S(Msg Msg:MsgType Failure Msg:MsgSubType Agent Msg:FromName AirCavityIDRedesigner Msg:FromType Redesigner Msg:Message "Decrease not possible in redesigner" Msg:Explanation (DECREASE HeadAirCavityID BY 1.e-3) Msg:ToName NIL Msg:ToType NIL Msg:Destination NIL Msg:InPlan HeadDP1 Msg:InMode Redesign Msg:Attitude Normal Msg:Constraints NIL Msg:SuggestionsBelow NIL Msg:Suggestions NIL Msg:ContributingMsg NIL)

```
-------------------- Leaving Redesigner
            ....AirCavityIDRedesigner
            ...Result=
```

#S(Msg Msg:MsgType Failure Msg:MsgSubType Agent Msg:FromName AirCavityIDRedesigner Msg:FromType Redesigner Msg:Message "Redesigner action section fails" Msg:Explanation (DECREASE HeadAirCavityID BY 1.e-3) Msg:ToName NIL Msg:ToType NIL Msg:Destination NIL Msg:InPlan HeadDP1 Msg:InMode Redesign ......

```
------------------ Leaving Step
            ....AirCavityID
            ...Result=
```

#S(Msg Msg:MsgType Failure Msg:MsgSubType Agent Msg:FromName AirCavityIDRedesigner Msg:FromType Redesigner Msg:Message "Redesigner action section fails" Msg:Explanation (DECREASE HeadAirCavityID BY 1.e-3) Msg:ToName NIL Msg:ToType NIL Msg:Destination NIL Msg:InPlan HeadDP1 Msg:InMode Redesign ......

!!! Note:
Now try the remaining suggestion.

```
------------------ Entering Step
            ...AirCavityDepth
    Redesign request from task :  AirCavity

-------------------- Entering Redesigner
            ...AirCavityDepthRedesigner
    Step = AirCavityDepth
    Suggest = (INCREASE HeadAirCavityDepth
                BY 1.e-3)

-------------------- Leaving Redesigner
            ....AirCavityDepthRedesigner
            ...Result= Success Msg
```

!!! Note:

This time the redesign works!

```
------------------ Leaving Step
            ....AirCavityDepth
            ...Result= Success Msg
```

!!! Note:
Finally a suggested backup step succeeds.
As none of the intermediate steps are affected
by the change in the Depth, the failing step
can be tried again under control of the
task's redesigner.

```
------------------ Entering Step
            ...CheckAirCavity
```

!!! Note:
Let us suppose that now the volume
is better and can be accepted.

???? ASK-USER: Head Air Cavity volume is 0.324
Is this OK for this design?
Answer is ????>yes

```
------------------ Leaving Step
            ....CheckAirCavity
            ...Result= Success Msg

----------------- Leaving Redesigner
            ....DefaultTaskRedesigner
            ...Result= Success Msg
```

!!! Note:
Now that everything has been patched
up the task's redesigner can quit.
The success is propagated back through
the failure handlers until the task
learns of the success. It can then
succeed itself.

```
---------------- Leaving FailureHandler
            ....DefaultUserStepFailureFH
            ...Result= Success Msg

-------------- Leaving FailureHandler
            ....SystemTaskBodyFailureFH
            ...Result= Success Msg
```

```
----------- Leaving Task
      ....AirCavity
      ...Result= Success Msg

   !!! etc !!!

--------- Leaving Plan
      ....HeadDP1
      ...Result= Success Msg

------- Leaving Specialist
      ....Head
      ...Result= Success Msg

   !!! etc !!!

----- Leaving Plan
     ....AirCylinderDP1
    ...Result= Success Msg

--- Leaving Specialist
    ....AirCylinder
   ...Result= Success Msg

*** Design attempt succeeds
*** Version date: (4 26 84)
*** Todays date:  (5 1 84)
*** User: DCBROWN
***** AIR-CYL Air-cylinder Design System *****
```

Appendix D: Plan Selection Trace

This is a trace generated by the system. Much of it has been edited out for brevity and presentation in this format. The trace was turned on to show plan selection. This trace is not from the Air-cylinder domain. This trace shows an example of a specialist SpA selecting amongst four plans P1, P2, P3 and P4. Each plan has two tasks selected from T1, T2 and T3. Each plan contains T1. The tasks each check one constraint. T1 always fails, while the others succeed. The selector and the sponsors are as follows:

```
(SPONSOR
  (NAME P1Sponsor)
  (USED-BY SpASelector)
  (PLAN P1)
  (COMMENT "each plan has a sponsor")
  (COMMENT "it returns a suitability")
  (BODY
   REPLY   (IF (ALREADY-TRIED? PLAN)
         THEN RULE-OUT)
   COMMENT "see if plan already tried"
   COMMENT "look for task T1 failing
        in last plan"
   REPLY   (IF (EQUAL 'T1
           LAST-FAILING-ITEM)
         THEN DONT-KNOW
         ELSE SUITABLE
         )
)))
(SPONSOR
  (NAME P2Sponsor)
  (USED-BY SpASelector)
  (PLAN P2)
  (BODY
   COMMENT  "plan already failed?"
   REPLY    (IF (ALREADY-TRIED? PLAN)
         THEN RULE-OUT)
   COMMENT  "Plan P3 failed already?"
   REPLY    (IF (ALREADY-TRIED? 'P3)
         THEN UNSUITABLE)
   COMMENT  "use requirements"
```

```
      Qualities (TABLE
            (DEPENDING-ON
              (RELIABILITY-REQS)
              (MANUFACTURABILITY-REQS)
              (COST-REQS)
            )
            (MATCH
             (IF (Reliable Easy Cheap)
              THEN PERFECT)
             (IF (Reliable ?    ? )
              THEN SUITABLE)
            )
            (OTHERWISE RULE-OUT)
            )
      COMMENT  "variable Qualities has a
               suitability as a value"
      COMMENT  "get a value from ddb"
      OpMed    (KB-FETCH
               'Requirements 'OperatingMedium)
      COMMENT  "now combine all info"
      REPLY    (TABLE
            (DEPENDING-ON
               OpMed  Qualities)
            (MATCH
             (IF ( Air    SUITABLE)
              THEN SUITABLE)
             (IF ( Air      ? )
              THEN DONT-KNOW)
            ))
) )

(SPONSOR
 (NAME P3Sponsor)
 (USED-BY SpASelector)
 (PLAN P3)
 (COMMENT "perfect unless
          already failed")
 (BODY
  REPLY    (IF (ALREADY-TRIED? PLAN)
           THEN RULE-OUT)
  REPLY    PERFECT
) )

(SPONSOR
 (NAME P4Sponsor)
 (USED-BY SpASelector)
 (PLAN P4)
 (COMMENT "perfect unless
          already failed")
```

```
    (BODY
     REPLY   (IF (ALREADY-TRIED? PLAN)
              THEN RULE-OUT)
     REPLY    PERFECT
) )

(SELECTOR
 (NAME SpASelector )
 (USED-BY SpA )
 (USES  P1Sponsor P2Sponsor
        P3Sponsor P4Sponsor)
 (TYPE  Design )
 (COMMENT "this is a selector")
 (COMMENT "it uses the output
           of the sponsors")
 (COMMENT "a s-mthd is a subroutine")
 (SELECTION-METHODS
   (METHOD (NAME M1)
       (INPUT-VARIABLE PlanNames)
       (BODY
         COMMENT "use designer preferred
                  order"
         REPLY (IF (ONLY-ONE? PlanNames)
              THEN PlanNames
                ELSE (DESIGNER-PREFERENCE
                      PlanNames)
) )    )   )
 (COMMENT "this is the start of the selection")
 (COMMENT "if plan P4 is perfect then use it")
 (COMMENT "otherwise use subroutine on
           perfect plans then suitable")
 (BODY
   REPLY  (IF (MEMBER 'P4 PERFECT-PLANS)
         THEN 'P4
         ELSE
          (IF PERFECT-PLANS
           THEN (USE-METHOD M1
                   ON PERFECT-PLANS)
           ELSE
            (IF SUITABLE-PLANS
             THEN (USE-METHOD M1
                     ON SUITABLE-PLANS)
             ELSE  NO-PLANS-APPLICABLE
         ) ) )
) )
```

The trace of the system is as follows:

```
***** AIR-CYL Air-cylinder Design System *****
*** Version date: (7 4 84)
*** Todays date:  (7 7 84)
*** User: DCBROWN


*** Requirements input
The following options are available:
   1 --- to use a set of standard test/demo requirements
   2 --- to have requirements read from your disk file
   3 --- to type them all in yourself
Note: you will be able to make alterations
*** Please type the number of your option >>>????>1

* Standard test/demo requirements to be used
     From file DCB:AC-Requirements-Test.LSP
Requirements:001

* Do you wish to make alterations to the requirements?
     Please answer YES or NO or QUIT >>>????>n

* No alterations to be made
*** Requirements Input Complete



--- Entering Specialist
    ...SpA... Mode = Design

!!! Note:
The specialist needs a plan.
Ask the sponsors for opinions
about each plan.

----- Entering Sponsor
     ...P1Sponsor...Plan = P1

----- Leaving Sponsor
     ....P1Sponsor...Result= SUITABLE

----- Entering Sponsor
     ...P2Sponsor...Plan = P2

----- Leaving Sponsor
     ....P2Sponsor...Result= SUITABLE

----- Entering Sponsor
     ...P3Sponsor...Plan = P3
```

----- Leaving Sponsor
....P3Sponsor...Result= PERFECT

----- Entering Sponsor
...P4Sponsor...Plan = P4

----- Leaving Sponsor
....P4Sponsor...Result= PERFECT

!!! Note:
Two suitable plans,
and two perfect.
The selector will pick one.

----- Entering Selector
...SpASelector

----- Leaving Selector
....SpASelector...Result= P4

!!! Note:
As directed, P4 was perfect
so it was picked.

----- Entering Plan
...P4... Type = Design

------- Entering Task...T1

!!! Note:
Plan 4 uses Task 1 which uses
Constraint 1 which has been
fixed to always fail.

--------- Entering TEST-CONSTRAINTS...(C1)

--------- Leaving TEST-CONSTRAINTS....(C1)
    ...Result=

#S(Msg Msg:MsgType Failure Msg:MsgSubType Constraint Msg:FromName C1 Msg:FromType Constraint Msg:Message "Constraint failure" Msg:Explanation "C1 forced failure" Msg:ToName NIL Msg:ToType NIL Msg:Destination NIL Msg:InPlan P4 Msg:InMode Design ......

!!! etc !!!

------- Leaving Task....T1
    ...Result=

#S(Msg Msg:MsgType Failure Msg:MsgSubType Agent Msg:FromName T1 Msg:FromType Task Msg:Message "Task failure" Msg:Explanation #S(Msg Msg:MsgType Failure Msg:MsgSubType FailureHandler Msg:FromName SystemTBFailureFH Msg:FromType FailureHandler ....

----- Leaving Plan....P4
   ...Result=

#S(Msg Msg:MsgType Failure Msg:MsgSubType Agent Msg:FromName P4 Msg:FromType Plan Msg:Message "Plan failure" Msg:Explanation NIL Msg:ToName NIL Msg:ToType NIL Msg:Destination NIL Msg:InPlan P4 Msg:InMode Design ......

!!! Note:
These agents know nothing about
redesign. The task fails and the
plan failure follows. A new plan
must be found. Ask the sponsors.

----- Entering Sponsor
   ...P1Sponsor...Plan = P1

----- Leaving Sponsor
   ....P1Sponsor...Result= DONT-KNOW

----- Entering Sponsor
   ...P2Sponsor...Plan = P2

----- Leaving Sponsor
   ....P2Sponsor...Result= SUITABLE

----- Entering Sponsor
   ...P3Sponsor...Plan = P3

----- Leaving Sponsor
   ....P3Sponsor...Result= PERFECT

----- Entering Sponsor
   ...P4Sponsor...Plan = P4

----- Leaving Sponsor
   ....P4Sponsor...Result= RULE-OUT

!!! Note:
Dont-know, suitable, perfect, rule-out
is the result.

----- Entering Selector
   ...SpASelector

----- Leaving Selector
    ....SpASelector...Result= P3

!!! Note:
The selector picks the perfect plan.

----- Entering Plan...P3... Type = Design

------- Entering Task...T3

--------- Entering TEST-CONSTRAINTS...(C2)

--------- Leaving TEST-CONSTRAINTS....(C2)
    ...Result= Success Msg

------- Leaving Task....T3
    ...Result= Success Msg

------- Entering Task...T1

!!! Note:
C1 fails again, leading to
plan failure.

--------- Entering TEST-CONSTRAINTS...(C1)

--------- Leaving TEST-CONSTRAINTS....(C1)
    ...Result=

#S(Msg Msg:MsgType Failure Msg:MsgSubType Constraint Msg:FromName C1 Msg:FromType Constraint Msg:Message "Constraint failure" Msg:Explanation "C1 forced failure" Msg:ToName NIL Msg:ToType NIL Msg:Destination NIL Msg:InPlan P3 Msg:InMode .....

    !!! etc !!!

------- Leaving Task....T1
    ...Result=

#S(Msg Msg:MsgType Failure Msg:MsgSubType Agent Msg:FromName T1 Msg:FromType Task Msg:Message "Task failure" Msg:Explanation #S(Msg Msg:MsgType Failure Msg:MsgSubType FailureHandler Msg:FromName SystemTBFailureFH Msg:FromType FailureHandler ....

----- Leaving Plan....P3
    ...Result=

#S(Msg Msg:MsgType Failure Msg:MsgSubType Agent Msg:FromName P3 Msg:FromType Plan Msg:Message "Plan failure" Msg:Explanation NIL Msg:ToName

NIL Msg:ToType NIL Msg:Destination NIL Msg:InPlan P3 Msg:InMode Design ......

!!! Note:
Ask the sponsors again.
Last time the result was
Dont-know, suitable, perfect, rule-out.

----- Entering Sponsor
   ...P1Sponsor...Plan = P1

----- Leaving Sponsor
   ....P1Sponsor...Result= DONT-KNOW

----- Entering Sponsor
   ...P2Sponsor...Plan = P2

----- Leaving Sponsor
   ....P2Sponsor...Result= UNSUITABLE

----- Entering Sponsor
   ...P3Sponsor...Plan = P3

----- Leaving Sponsor
   ....P3Sponsor...Result= RULE-OUT

----- Entering Sponsor
   ...P4Sponsor...Plan = P4

----- Leaving Sponsor
   ....P4Sponsor...Result= RULE-OUT

!!! Note:
The result is
dont-know, unsuitable, rule-out, rule-out.
The selector gives up.

----- Entering Selector
   ...SpASelector

----- Leaving Selector....SpASelector
   ...Result= ApplicablePlanNotFound

!!! Note:
With no more plans to try
the specialist will fail.

--- Leaving Specialist....SpA
  ...Result=

#S(Msg Msg:MsgType Failure Msg:MsgSubType Agent Msg:FromName SpA Msg:FromType Specialist Msg:Message "Applicable plan not found" Msg:Explanation NIL Msg:ToName NIL Msg:ToType NIL Msg:Destination NIL Msg:InPlan TOP Msg:InMode Design Msg:Attitude Normal Msg:Constraints NIL Msg:SuggestionsBelow NIL Msg:Suggestions NIL Msg:ContributingMsg NIL)

*** Design attempt fails
*** Version date: (7 4 84)
*** Todays date: (7 7 84)
*** User: DCBROWN
***** AIR-CYL Air-cylinder Design System *****

Appendix E: DSPL Syntax

We will use examples to show the variety of the syntax in the design language. The agents used as examples will not be actual AIR-CYL samples but constructed examples designed to show the various allowable constructs and their uses.

KEY:
- Nm   Name
- Sp   Specialist
- T   Task
- S   Step
- FH   Failure Handler
- P   Plan
- PSp Plan Sponsor
- PSe Plan Selector
- C   Constraint
- Q   Quality (eg. Cheap, Reliable, Light, ... ANYTHING)
- Fr   Frame in Design Data Base
- RD   Redesigner
- A   Attribute

```
(SPECIALIST
 (NAME SpNm)
 (USED-BY SpNm)
 (USES SpNm SpNm SpNm)
 (COMMENT "some comment")
 (DESIGN-PLANS PNm PNm)
 (DESIGN-PLAN-SELECTOR  PSe)
 (ROUGH-DESIGN-PLANS PNm)
 (ROUGH-DESIGN-PLAN-SELECTOR PSe)
 (INITIAL-CONSTRAINTS CNm CNm)
 (FINAL-CONSTRAINTS CNm)
)


(SELECTOR
 (NAME PSeNm)
 (USED-BY SpNm)
 (USES PSpNm PSpNm)
 (TYPE Design )
```

```
(SELECTION-METHODS
 (METHOD
  (NAME M1)
  (INPUT-VARIABLE x)
  (BODY
   REPLY (IF (ONLY-ONE? x)
        THEN x
        ELSE (DESIGNER-PREFERENCE x)
))) )
(BODY
 REPLY
  (IF (MEMBER 'PNm PERFECT-PLANS)
   THEN 'PNm
   ELSE
    (IF PERFECT-PLANS
     THEN (USE-METHOD M1 ON PERFECT-PLANS)
     ELSE
      (IF SUITABLE-PLANS
       THEN (USE-METHOD M1 ON SUITABLE-PLANS)
       ELSE NO-PLANS-APPLICABLE
)) ) ) )


(SPONSOR
 (NAME PSpNm)
 (USED-BY PSeNm)
 (PLAN PNm)
 (BODY
  REPLY (IF (ALREADY-TRIED? PLAN) THEN RULE-OUT)
  REPLY (IF (ALREADY-TRIED? 'PNm) THEN UNSUITABLE)
  pp    (IF (EQUAL 'TNm LAST-FAILING-ITEM)
       THEN DONT-KNOW
       ELSE SUITABLE
       )
  qq    (TABLE
       (DEPENDING-ON (RELIABILITY-REQS)
              (MANUFACTURABILITY-REQS)
              (COST-REQS)
       )
       (MATCH
        (IF (Reliable Easy Cheap) THEN PERFECT)
        (IF (Reliable ?    ? ) THEN SUITABLE)
       )
       (OTHERWISE RULE-OUT)
       )
  rr    (KB-FETCH 'Requirements 'OperatingMedium)
  REPLY (TABLE
       (DEPENDING-ON
```

```
                pp    rr    qq)
       (MATCH
         (IF ( SUITABLE  Air   SUITABLE)
          THEN SUITABLE)
         (IF ( DONT-KNOW  Air    ?  )
          THEN DONT-KNOW)
)) ))


(PLAN
 (NAME PNm)
 (TYPE Design)
 (USED-BY SpNm)
 (USES SpNm SpNm SpNm)
 (SPONSOR SponsorNm)
 (COMMENT "some comment")
 (INITIAL-CONSTRAINTS CNm)
 (FINAL-CONSTRAINTS CNm CNm)
 (BODY TNm
      (ROUGH-DESIGN SpNm)
      (DESIGN SpNm)
      (TEST-CONSTRAINT CNm)
      (PARALLEL (DESIGN SpNm)
            (DESIGN SpNm)
      )
      (REPORT-ON FrNm)
))


(PLAN
 (NAME PNm)
 (TYPE RoughDesign)
 (USED-BY SpNm)
 (USES SpNm SpNm SpNM)
 (SPONSOR SponsorNm)
 (INITIAL-CONSTRAINTS None)
 (FINAL-CONSTRAINTS CNm)
 (BODY
  (ROUGH-DESIGN SpNm)
  (TEST-CONSTRAINT CNm)
  TNm
  (ROUGH-DESIGN SpNm)
))


(TASK
 (NAME TNm)
 (USED-BY PNm)
```

```
  (INITIAL-CONSTRAINTS None)
  (FINAL-CONSTRAINTS None)
  (FAILURE-HANDLER
   USER-STEP-FAILURE-FH IS FHNm
  )
  (FAILURE-SUGGESTIONS
   (SUGGEST (DECREASE ANm))
   (SUGGEST (INCREASE ANm))
  )
  (REDESIGNER  RDNm)
  (BODY
   SNm
   SNm
   (TEST-CONSTRAINT CNm)
   SNm
))


(TASK
 (NAME TNm)
 (USED-BY PNm)
 (INITIAL-CONSTRAINTS CNm)
 (FINAL-CONSTRAINTS None)
 (REDESIGN  NOT-POSSIBLE)
 (FAILURE-SUGGESTIONS
  (SUGGEST (DECREASE ANm))
 )
 (COMMENT "some comment")
 (BODY
  SNm
  SNm
))


(STEP
 (NAME SNm)
 (USED-BY TNm TNm)
 (ATTRIBUTE-NAME ANm)
 (REDESIGNER  RDNm)
 (FAILURE-SUGGESTIONS
  (SUGGEST (DECREASE ANm))
  (SUGGEST (DECREASE ANm))
  (SUGGEST (INCREASE ANm))
  (SUGGEST (CHANGE PistonMaterial
            TO INCREASE MinThickness))
 )
 (COMMENT "a design step")
 (BODY
```

```
(KNOWN
 Material (KB-FETCH 'aaa 'Material)
 MinTh   (KB-FETCH Material 'MinThickness)
 xxx     0.021
 yyy     (LNGTH 1.234 0.01 0.001 'ThreeDP)
 )
(DECISIONS
 zzz    (IF ( >= MinTh yyy)
        THEN (- yyy
              (+ MinTh xxx)
             )
        ELSE (- yyy
              (DOUBLE MinTh)
             ) )
 ppp    (HALF (- xxx zzz))
 REPLY  (TEST-CONSTRAINT CNm)
 qqq    (THREE-DP
         (+ zzz (LARGER ppp MinTh))
        )
 REPLY  (KB-STORE 'aaa 'ANm)
)))
```

Available:
 (+ a b) (- a b)
 (* a b) (// a b) (^ a 2)
 (SQRT a) (DOUBLE a) (HALF a)
 (LARGER a b) (SMALLER a b)
 (VALUE a) (VALUE- a) (VALUE+ a)
 (+TOLERANCE a) (-TOLERANCE a)
 (MAX-TOLERANCE a)
 (THREE-DP a) (TWO-DP a)
 (STANDARDIZE-UP a)
 (STANDARDIZE-DOWN a)
 (STANDARDIZE-NEAREST a)
 (ASK-VALUE ....)
 (ASK-USER ....)
 (= a b) (~= a b) (> a b) (< a b)
 (>= a b) (<= a b) (>=< a b)
 (IS a b) (IS-ONE-OF a '(b c d))

```
(STEP
 (NAME SNm)
 (USED-BY TNm)
 (COMMENT "Checking Requirements")
 (BODY
```

```
  (KNOWN
   lll  (KB-FETCH 'Requirements 'lll)
   hhh  (KB-FETCH 'Requirements 'hhh)
   ppp  (KB-FETCH 'Requirements 'ppp)
  )
  (DECISIONS
    REPLY (IF (OR (= ppp 1.5)(= ppp 1.75))
         THEN OK
         ELSE (FAILURE "message")
         )
    REPLY (TABLE
         (DEPENDING-ON
               lll    hhh
         )
         (MATCH
          (IF (  (<= 6)   (<= 2) )
           THEN (FAILURE "failure message"))
          (IF ( (6 >=< 7) (1 >=< 2)) THEN OK)
          (IF ( (8 >=< 9) (2 >=< 3)) THEN OK)
         )
         (OTHERWISE (FAILURE "message"))
         )
)))


(STEP
 (NAME SNm)
 (USED-BY TNm)
 (ATTRIBUTE-NAME ANm)
 (COMMENT "rough design")
 (BODY
  (KNOWN
   fff (KB-FETCH 'Requirements 'fff)
   ggg (KB-FETCH 'Requirements 'ggg)
   hhh (KB-FETCH 'zzz 'hhh)
  )
  (DECISIONS
   jjj  (TABLE
        (DEPENDING-ON
              fff   ggg    hhh
        )
        (MATCH
         (IF ( (<= 250) Reliable Corrosive )
          THEN StainlessSteel)
         (IF ( (<= 200)   ?   NonCorrosive )
          THEN Aluminum)
        )
        (OTHERWISE StainlessSteel)
```

```
    )
   REPLY (KB-STORE 'ddd ANm jjj)
)))


(CONSTRAINT
 (NAME CNm)
 (USED-BY  SNm SNm)
 (COMMENT "some comment")
 (FAILURE-MESSAGE  "message")
 (FAILURE-SUGGESTIONS
  (SUGGEST (DECREASE nnn BY FAILURE-AMOUNT))
  (SUGGEST (INCREASE kkk BY ArithExpr))
  (IF ( > aaa bbb)
   THEN
    (SUGGEST (DECREASE mmm BY FAILURE-AMOUNT))
  )
 )
 (BODY
  (TEST
   ( aaa  <=  (+ zzz (HALF bbb)) )
)))

(FAILURE-HANDLER
 (NAME FHNm )
 (TYPE System)
 (USED-BY-TYPE Step )
 (COMMENT "some comment")
 (BODY
  (TABLE
   (DEPENDING-ON  MESSAGE)
   (MATCH
    (IF ( "Programming problem" )
     THEN (FAIL))
    (IF ( "Use of plan language problem" )
     THEN (FAIL))
    (IF ( "some message" )
     THEN  (USE-FH SystemFHNm
              WITH CONTRIBUTING-MSG))
    (IF ( "some other message" )
     THEN  (USE-FH UserFHNm)
   ))
   (OTHERWISE
    (DO (COMPLAIN "Message not recognized")
        (FAIL)
   ))
)))
```

```
(FAILURE-HANDLER
 (NAME FHNm)
 (TYPE USER-DECISION-CONSTRAINT-FH)
 (USED-BY SNm)
 (USED-BY-TYPE Step)
 (COMMENT "written by user")
 (BODY
  (TABLE
   (DEPENDING-ON  EXPLANATION)
   (MATCH
    (IF ( "constraint failure message" )
     THEN
      (DO (COMPLAIN "Redesign here !!")
         (STEP-REDESIGN-WITH-SUGGESTIONS)
      )
   ))
   (OTHERWISE
    (USE-FH SystemConstraintFH
         WITH CONTRIBUTING-MSG)
   )
)))

(FAILURE-HANDLER
 (NAME FHNm)
 (TYPE USER-STEP-FAILURE-FH)
 (USED-BY TNm)
 (USED-BY-TYPE Task)
 (COMMENT "will default to this")
 (BODY
  (TASK-REDESIGN-WITH-SUGGESTIONS)
))


(REDESIGNER
 (NAME RDNm)
 (USED-BY  SNm)
 (USED-BY-TYPE  Step)
 (ADJUSTMENT  0.001)
 (VALUE  (KB-FETCH 'xxx 'ANm))
 (COMMENT "some comment")
 (INCREASE
  (KNOWNS
   ppp      (KB-FETCH 'ccc 'ppp)
   Material (KB-FETCH 'uuu 'Material)
   MinTh    (KB-FETCH Material 'MinThickness)
   sss      (KB-FETCH 'ccc 'qqq)
  )
  (DECISION
```

```
    sss   (IF ( > sss MinTh)
          THEN (- ppp
                (+ ppp MinTh)
               )
          ELSE (- ppp
                (DOUBLE MinTh)
          )  )
   www   (+ VALUE INCREASE)
   REPLY (TEST-CONSTRAINT CNm)
   REPLY (KB-STORE 'ccc 'ANm www)
 ))
 (DECREASE
  (DECISION
   www   (- VALUE DECREASE)
   REPLY (IF (< www 0.1)
         THEN DECREASE-NOT-POSSIBLE
         ELSE (KB-STORE 'ccc 'ANm www)
         )
 ))
 (CHANGE  CHANGE-NOT-POSSIBLE)
)
```